AF806310

Studies in European Economic Law and Regulation

Volume 22

Series Editors

Kai Purnhagen, University of Bayreuth, Bayreuth, Germany

Josephine van Zeben, Wageningen University & Research, Wageningen, The Netherlands

Editorial Board Members

Alberto Alemanno, HEC Paris, Paris, France

Mads Andenaes, University of Oslo, Oslo, Norway

Stefania Baroncelli, University of Bozen, Bozen, Italy

Franziska Boehm, Westfälische Wilhelms-University Münster, Münster, Germany

Anu Bradford, Columbia Law School, New York, USA

Jan Dalhuisen, King's College London, London, UK

Michael Faure, Maastricht University, Maastricht, The Netherlands

Jens-Uwe Franck, Ludwig-Maximilians-University Munich, Munich, Germany

Geneviève Helleringer, University of Oxford, Oxford, UK

Christopher Hodges, University of Oxford, Oxford, UK

Lars Hornuf, University of Bremen, Bremen, Germany

Moritz Jesse, Leiden University, Leiden, The Netherlands

Marco Loos, University of Amsterdam, Amsterdam, The Netherlands

Petros Mavroidis, Columbia Law School, New York, USA

Hans Micklitz, European University Institute, Florence, Italy

Giorgio Monti, European University Institute, Florence, Italy

Florian Möslein, Philipps-University of Marburg, Marburg, Germany

Dennis Patterson, European University Institute, Florence, Italy

Wolf-Georg Ringe, University of Hamburg, Hamburg, Germany

Jules Stuyck, Katholieke Universiteit Leuven, Leuven, Belgium

Bart van Vooren, University of Copenhagen, Copenhagen, Denmark

The series "Studies in European Economic Law and Regulation" is devoted to the analysis of European Economic Law. The series' scope covers a broad range of topics within economics law including, but not limited to, the relationship between EU law and WTO law; free movement under EU law and its impact on fundamental rights; antitrust law; trade law; unfair competition law; financial market law; consumer law; food law; and health law. These subjects are approached both from doctrinal and interdisciplinary perspectives.

The series accepts monographs focusing on a specific topic, as well as edited collections of articles covering a specific theme or collections of articles. All contributions are subject to rigorous double-blind peer-review.

More information about this series at https://link.springer.com/bookseries/11710

Eva van der Zee

Sustainability Labels in the Shadow of the Law

Springer

Eva van der Zee
Institute of Law and Economics
University of Hamburg
Hamburg, Germany

ISSN 2214-2037 ISSN 2214-2045 (electronic)
Studies in European Economic Law and Regulation
ISBN 978-3-030-95801-5 ISBN 978-3-030-95802-2 (eBook)
https://doi.org/10.1007/978-3-030-95802-2

© The Editor(s) (if applicable) and The Author(s), under exclusive license to Springer Nature Switzerland AG 2022
This work is subject to copyright. All rights are solely and exclusively licensed by the Publisher, whether the whole or part of the material is concerned, specifically the rights of translation, reprinting, reuse of illustrations, recitation, broadcasting, reproduction on microfilms or in any other physical way, and transmission or information storage and retrieval, electronic adaptation, computer software, or by similar or dissimilar methodology now known or hereafter developed.
The use of general descriptive names, registered names, trademarks, service marks, etc. in this publication does not imply, even in the absence of a specific statement, that such names are exempt from the relevant protective laws and regulations and therefore free for general use.
The publisher, the authors and the editors are safe to assume that the advice and information in this book are believed to be true and accurate at the date of publication. Neither the publisher nor the authors or the editors give a warranty, expressed or implied, with respect to the material contained herein or for any errors or omissions that may have been made. The publisher remains neutral with regard to jurisdictional claims in published maps and institutional affiliations.

This Springer imprint is published by the registered company Springer Nature Switzerland AG
The registered company address is: Gewerbestrasse 11, 6330 Cham, Switzerland

For Bart and Willeke

Foreword

What's in a Label?

When we shop, we are exposed to food labels. The credence character of foodstuffs requires consumers to rely on the information contained by the respective label attached to food. But not only consumers need accessible, reliable, and encompassing information on the label to assess the value of the foodstuff. Along the supply chain, producers, retailers, and processing companies rely on the information of the food provided to them. The important role information plays in the food market has several effects: In B2B relationships, businesses compete to provide their clients with the best available information to acquire a comparative advantage over other businesses on the market. In B2C relationships, consumers trust products with specific on-pack information, such as a sustainability label, more than products without a label. That also opens opportunities for businesses to gain a comparative advantage by attaching a label to the product. Dr. van der Zee shows illustratively in Chap. 2 how many private labeling schemes exist, what is their regulatory structure, and how these can be used to increase consumer confidence. Unlike in B2B relationships, where professionals act in the chain, in B2C relationships consumers do not have the means nor the knowledge available to check for the reliability of the respective information. While for this reason on-pack information is in general heavily regulated in the EU, voluntary labels do not enjoy a comparable set of regulations. Using surveys Dr. van der Zee illustrates how consumers can be systematically misled into buying a foodstuff even if the label attached to the food is substantially meaningless. By doing that she justifies to our knowledge for the first time empirically that such pictograms would require regulation to protect consumers. But it is not only the consumer which needs protection by the law. Businesses, which may also want to control the image of their product via information, may be protected by freedom of speech rights. If these exist and to what extent they exist is subject to analysis by Dr. van der Zee's Chap. 4. Private standards are often the only effective regulation of global food supply chains. This provides standard setters

with a powerful position, namely controlling access to the respective supply chain. Consequently, Dr. van der Zee turns to evaluating whether private standards are or should be under scrutiny of EU competition law according to Art. 101 TFEU. After evaluating the state of the art, she introduces a human well-being test, which opens new possibilities to evaluate private standards under competition law.

The book provides a thorough, well-researched, and innovative approach to the regulation of private food standards. It is innovative as it uses a variety of methods, embracing doctrinal, empirical, and philosophical methods. The solutions presented are likewise innovative and go way beyond the state of the art. Yet, they are well-reasoned and grounded on solid research. All in all, a book worth reading for scholars and practitioners alike.

University of Bayreuth,
Bayreuth, Germany
August 2021

Kai P. Purnhagen

Acknowledgments

This book would not have been possible had it not been for the most excellent guidance that I received as a PhD candidate at Wageningen University as well as during my visiting scholarships at the European University Institute and New York University. I would like to thank my supervision team for their wisdom, support, and guidance. Thank you Katrien Termeer, for your expert advice and encouragement. Thank you Laurence Gormley, for your sharp mind and extraordinary support. Thank you Arnout Fischer, for helping me transgress the boundaries of my own discipline. And last but certainly not least, thank you Kai Purnhagen, for being an inspirational guide in (multidisciplinary) legal research. Thank you all for always taking the time to look at my work and providing me with helpful comments. I would also like to thank Omri Ben-Shahar, Marco Bronckers, Marise Cremona, Gráinne de Búrca, Simon Deakin, Eleanor M. Fox, Robert Howse, Hans Micklitz, Giorgio Monti, Peter Oosterveer, Joanne Scott, and Jules Stuyck for taking the time to discuss my work and for your guidance. A big thank you to all my dear friends and family for your love and support.

Contents

1 Introduction 1

2 Regulatory Structure Underlying Sustainability Labels 17

3 Sustainability Labels in the Shadow of EU Consumer Law 33

4 Sustainability Labels in the Shadow of EU Fundamental Rights Law 51

5 Sustainability Labels in the Shadow of EU Competition Law 81

6 Conclusion and Discussion 101

Index 119

About the Author

Eva van der Zee is a Junior Professor (Tenure Track) at the Institute of Law and Economics, Hamburg University. Her main research area is in international and EU law with a focus on behavioral law and economics. Before being appointed at Hamburg University, Eva van der Zee was a Postdoctoral Researcher at the Strategic Communication Group at Wageningen University and a PhD Researcher at the Law and Governance Group and the Marketing and Consumer Behavioral Group at Wageningen University. She has been a visiting researcher at New York University and the European University Institute.

List of Abbreviations

ACC	Advertising Code Committee
ACM	Autoriteit Consument en Markt
AICDP	Agreement on the International Dolphin Conservation Program
ASC	Aquaculture Stewardship Council
CAC	Command and control
CBA	Cost-benefit analysis
CECED	European Committee of Domestic Equipment Manufacturers
Cf	Compare
CFR	Charter of Fundamental Rights of the European Union
d	Effect size
DAC	Dutch accreditation body
E.g.	For example
EC	European Communities
ECHR	European Convention on Human Rights
ECJ	European Court of Justice
ECSC	European Coal and Steel Community
ECtHR	European Court of Human Rights
EEC	European Economic Community
ETP	Eastern Tropical Pacific
EU	European Union
EVU	European Vegetarian Union
FDA	Food and Drug Administration
FIR	Food Information Regulation
GATT	General Agreement on Tariffs and Trade
GFL	General Food Law
GLOBALG.A.P.	Global good agricultural practice
GM	Genetically modified
HAP-LH	Human, animal, or plant life or health
i.e.	That is
ICJ	International Court of Justice
IEC	International Electrotechnical Commission

IGO	Intergovernmental organization
IIA	International investment agreement
ISEAL	International Social and Environmental Accreditation and Labelling
ISO	International Organization for Standardization
ITA	International trade agreement
KPMG	Klynveld, Peat, Marwick, and Goerdeler
LSD	Least significant difference
M	Mean
MSC	Marine Stewardship Council
n	Sample size
NCA	National competition authority
NGO	Non-governmental organization
p	Significance
PDO	Protected designation of origin
PGI	Protected geographical indication
PPM	Process and production method
SD	Standard deviation
SPS	Sanitary and phytosanitary measures
SWB	Subjective well-being
TBT	Technical barriers to trade
TEU	Treaty on European Union
TFEU	Treaty on the Functioning of the European Union
TSG	Traditional specialty guaranteed
UCPD	Unfair Commercial Practices Directive
UK	United Kingdom
UN	United Nations
USA	United States of America
VSS	voluntary sustainability standard
WTO	World Trade Organization
WTP	Willingness to pay

Chapter 1
Introduction

Sustainability sells. Sustainable food is gaining popularity and businesses are taking note. Many food products are now labelled with sustainability information through logos, symbols, and texts.[1] Such business-to-consumer (B2C) sustainability labels may reduce information asymmetry if they adequately communicate sustainability attributes to consumers. Consumers require such adequate communication since producers of food products often have more or better information about specific sustainability attributes of products or production processes compared to consumers. Information asymmetry is especially apparent for sustainability attributes, as these can be considered credence attributes[2] and Potemkin attributes.[3] This is less so for search attributes that are known before purchase and experience attributes that are known costlessly only after purchase.[4] Credence attributes, such as the use of pollutants in the production process, are expensive to determine even after purchase,[5] while Potemkin attributes, such as fair trade and animal welfare, cannot be verified by the consumer or external institutions at the end-product level.[6]

However, sustainability labels may not always be effective in communicating sustainability attributes to consumers. Because shopping is often a hurried affair where different information cues compete for attention, consumers have difficulty processing communications on food products.[7] As such, consumers may not pick up on sustainability labels. However, even if consumers pick up on sustainability labels they may fail to select a reliable one, because food businesses may apply sustainability labels to unsustainable products to secure a price premium, known as

[1] Hahnel et al. (2015), p. 1392; For a definition of a label see Codex, 1985 rev. 1991.

[2] Grunert et al. (2014), p. 177, Chap. 2, n 1; Van Dam and Van Trijp (2011), p. 726.

[3] Spadoni et al. (2014), p. 142.

[4] Nelson (1970), p. 311.

[5] Darby and Karni (1973), p. 67.

[6] Spadoni et al. (2014), n 3.

[7] Horne (2009), p. 175, Chap. 2, n 5. See also Becher and Feldman (2016).

© The Author(s), under exclusive license to Springer Nature Switzerland AG 2022

E. van der Zee, *Sustainability Labels in the Shadow of the Law*, Studies in European Economic Law and Regulation 22, https://doi.org/10.1007/978-3-030-95802-2_1

greenwashing.[8] Such greenwashing practices increase the multiplicity and diversity of sustainability labels with different degrees of reliability, credibility and effectiveness on the consumer market, which will only make it harder for consumers to pick up on reliable sustainability labels. This is problematic because such information asymmetry between consumers and producers can impede the smooth functioning of markets, which in the worst case may lead to market failure. Market failure implies that consumers are unwilling to pay more for better quality, leading to a lack of motivation by producers to offer higher quality products.[9]

This book describes and examines three EU legal frameworks (EU competition law, EU consumer law, and EU fundamental rights law) that may affect the extent to which consumers consume more sustainably, i.e. opting for a more sustainable option compared to the conventional product. In doing so, this book goes beyond a rationalist interpretation derived from neoclassical theory on the interpretation and application of EU law. Neoclassical theory has severely impacted the interpretation and application of EU law.[10] Neoclassical theory assumes that individuals act as "rational preference-maximizers who respond to incentives".[11] Law is then considered within neoclassical theory as a price that shapes such incentives.[12] It is assumed that when an individual, the *homo economicus*, is confronted with a variety of choices it will choose the option that yields it the most expected economic welfare.[13] This *homo economicus*, is assumed to be aware of all its preferences, knows all ins and outs of the options with which it is presented and is perfectly able to choose the option that maximizes its own welfare.[14]

Practice shows that the implications of using the noncritical application of these normative assumptions in the interpretation and application of EU competition law, EU consumer law, and EU fundamental rights law to sustainability labels may have an inhibiting effect on sustainable consumption by amplifying the opportunity of market failure. When the interpretation and application of EU law based on neoclassical assumptions have such an inhibiting effect can be illustrated by three examples from EU competition law, EU consumer law, and EU fundamental rights law.

A first example is the practice by the Dutch Competition Authority (ACM) who applied Article 101 TFEU (Treaty on the Functioning of the EU) in such a way that a private sustainability initiative, known as the Chicken of Tomorrow, was interpreted

[8] Cafaggi and Iamiceli (2013), p. 131, Chap. 3, n 2.

[9] Akerlof (1970), p. 488.

[10] See on the impact of neoclassical law and economics in EU unfair commercial practices law: Micklitz (2008); Faure and Luth (2011). See on the impact of neoclassical law and economics in EU competition law: Heinemann (2015).

[11] Nourse and Shaffer (2009), p. 66.

[12] Ibid, 66.

[13] For a summary and defence of rational choice theory, see the founding father of law and economics, Posner (1997), p. 1551.

[14] Faure and Luth (2011), n 10.

to be an anti-competitive agreement, which could not be justified as consumers did not receive a fair share of the resulting benefits.[15] Chicken of Tomorrow was developed in 2013 by Dutch supermarkets to collectively stop the sale of chicken meat that did not meet the criteria set by the Chicken of Tomorrow initiative. As such, the initiative may promote sustainable consumption since a less sustainable option will no longer be available on the market. Furthermore, it limits a multiplicity of sustainability labels that set lower requirements than those set by Chicken of Tomorrow. However, private sustainability initiatives such as Chicken of Tomorrow may also limit competition as they restrict market access to producers of conventional chicken meat. Depending on whether such an initiative can be considered an agreement that limits competition, and how consumer fair share of such an agreement is interpreted under EU competition law, affects whether private sustainability initiatives that stop the sale of products that do not meet certain sustainability criteria would be prohibited under Article 101 TFEU.

The ACM seemed to have based its interpretation of 'consumer fair share' on the normative assumption of the rational actor derived from neoclassical theory. The ACM interpreted 'consumer fair share' as the economic welfare in monetary terms of the individual consumer of the product. To examine how much the individual consumer gained in economic welfare in monetary terms from the Chicken of Tomorrow agreement, the ACM used a classical cost-benefit analysis (CBA) based on rational choice theory. The ACM assessed the costs and benefits of private sustainability agreements, using shadow prices and consumer willingness to pay (WTP) to monetarize sustainability benefits. In doing so, the ACM based its assessment of 'consumer fair share' on the assumption of the consumer as a *homo economicus*, who would be perfectly capable to choose the option that would maximize their welfare in monetary terms. Based on the CBA, the ACM concluded that the increased price that the ACM assumed to result from the agreement, did not exceed the monetarized benefits to the individual consumer. The ACM advised the supermarkets to dissolve the agreement.

Crucially, the analysis of the ACM is explicitly publicised as a guide to companies to assess whether their private sustainability agreements comply with the EU competition rules.[16] However, the analysis by the ACM fails to accurately estimate the benefits of private sustainability initiatives as monetizing sustainable development benefits based on the idea that how much money an individual consumer is willing to pay or accept for increased sustainable production assumes that human beings are rational decision-makers, while human beings are often not rational but

[15] Autoriteit Consument en Markt, 'Analyse ACM van Duurzaamheidsafspraken 'De Kip van Morgen'' (*Autoriteit Consument en Markt,* 2015) <https://www.acm.nl/nl/download/publicatie/?id=13758>.

[16] Ibid; Monti and Mulder (2017), p. 641. A more recent draft guide can be found here: Autoriteit Consument & Markt, 'Guidelines Sustainability Agreements. Opportunities within Competition Law', https://www.acm.nl/sites/default/files/documents/second-draft-version-guidelines-on-sustainability-agreements-oppurtunities-within-competitionlaw.pdf.

subject to several biases.[17] As a consequence, the analysis by the ACM may discourage private sustainability initiatives that may otherwise not act due to the risks of free-riders (competitors free ride on the sustainability efforts—e.g. information campaigns—made by another company) and first-mover disadvantages (first company only selling sustainable products loses customers to competitors offering cheaper unsustainable alternatives),[18] thereby potentially reducing sustainable consumption.

A second example is the practice by the Dutch Advertising Code Committee (ACC)[19] regarding a private label stating "sustainable eel fund". This label does not require sustainable eel fishing, which would be an impossible requirement according to many environmental organizations,[20] but solely requires that the eel fisher contributes money to the fund. According to Greenpeace, this label was misleading, and it filed a complaint with the ACC. The ACC is a self-regulatory initiative by Dutch businesses to regulate advertisements, which offers a low-threshold forum for individuals to complain about misleading advertisements.[21] The ACC held that the label was not misleading as consumers would understand that by referring to a "fund" the label did not claim that the eel was sustainably fished.[22] To come to this decision, the ACC relied on the normative benchmark of the "average consumer" who is 'reasonably well-informed, reasonably observant and circumspect' as clarified by the European Court of Justice (ECJ).[23] The ACC interpreted the normative benchmark of the "average consumer" in such a way that the "average consumer" would understand the "sustainable eel fund" label.

It may well be that this interpretation of the "average consumer" benchmark is inadequate. Although the "average consumer" test is often, especially at the EU level, interpreted to reflect the *homo economicus,*[24] it is also often been criticized for overlooking real consumer behaviour.[25] Interpreting the "average consumer" as a *homo economicus* may have the effect that labels that give consumers the false impression of sustainability may not be prohibited. When many labels on the market give a false impression of sustainability, it may become difficult for consumers to

---

[17] The revolution in psychology that has changes perceptions of how people act in economic circumstances may have started with Herbert Simon's paper in 1955 for which he won the 1978 Nobel Prize in economics that introduced the concept of bounded rationality. Simon (1955), p. 99.

[18] Monti and Mulder (2017), n 16, p. 636.

[19] See for more information on the working procedures of the Advertising Code Committee: Stichting Reclame Code, 'The Dutch Advertising Code and Working Procedure' (Stichting Reclame Code, 2022), https://www.reclamecode.nl/nrc_taxonomy/general/?lang=en.

[20] Case Greenpeace v DUPAN *Greenpeace v DUPAN*, Dossier 2011/00590.

[21] See ibid.

[22] Ibid.

[23] See e.g., Case 315/92 *Clinique*, ECLI:EU:C:1994:34 and Case 210/96 *Gut Springheide*, ECLI:EU:C:1998:369.

[24] Incardona and Poncibò (2007), p. 21; Howells and Weatherill (2005).

[25] Franck and Purnhagen (2014); Incardona and Poncibò (2007); Schebesta and Purnhagen (2016), p. 590; Sibony and Helleringer (2015); ibid; Trzaskowski (2011), p. 383; Purnhagen (2017), p. 437.

distinguish the true from the false claims. As a consequence, free-rider problems may emerge where consumers purchase the cheaper conventional product thinking it is sustainable, which may outcompete the producer of the sustainable alternative. This may reduce sustainable consumption.

A third example is that businesses are mandated at the EU level to provide a disclaimer on genetically modified (GM) food products that the product—or certain ingredients—contains, consists of, or is obtained from genetic modification.[26] The objective of the Regulation is to facilitate "accurate labelling" to enable consumers "to exercise their freedom of choice in an effective manner".[27] In theory, this may promote sustainable consumption as GM crops are often considered to pose serious threats to farmer sovereignty and national food security.[28] A main concern is that most GM crops have been modified for herbicide tolerance. This has led to toxic herbicides, such as the widely discussed and protested Roundup herbicide, increasing tremendously since GM crops were first introduced. If consumers do not want to purchase GM foods, mandatory GM labelling could, hence, be considered to promote sustainable consumption. Information interventions, such as disclaimers, have long been an important policy tool to enhance consumers' ability to navigate the marketplace effectively. It is based on the assumption of the homo economicus that the availability of relevant, accurate information will assist consumers to make meaningful purchase decisions. Experimental studies from cognitive science and behavioural economics show, however, that mandatory disclaimers increase consumer confusion, and may often be ineffective or harmful to perceptions, attitudes, and decisions.[29] As such, a noncritical application of the rational actor model to mandatory disclaimers may lead to inaccurate labelling, potentially further reducing sustainable consumption. This is especially relevant in EU fundamental rights law, where mandatory disclaimers, if ineffective to inform consumers adequately, may potentially clash with the freedom of expression of food businesses.

These three examples illustrate that the interpretation and application of EU law based on the noncritical application of neoclassical assumptions may have an inhibiting effect on sustainable consumption by amplifying the opportunity of market failure through (1) discouraging private sustainability initiatives that ban unsustainable products, (2) allowing labels that give a false impression of sustainability, and (3) mandating disclaimers on sustainability attributes that consumers are not able to process effectively. As such, a complex situation has arisen where sustainable consumption and production is an important priority in EU law and

[26] Regulation (EC) No 1830/2003 of the European Parliament and of the Council of 22 September 2003 concerning the traceability and labelling of genetically modified organisms and the traceability of food and feed products produced from genetically modified organisms and amending Directive 2001/18/EC, Article 4B.

[27] Regulation (EC) No 1830/2003, Article 1 in conjunction with Recital 4.

[28] Zerbe (2004), p. 593.

[29] Green and Scott Armstrong (2012), p. 293.

policy,[30] while the interpretation and application of EU law may ultimately inhibit such sustainable consumption by stimulating greenwashing on food labels and discouraging best practices in sustainable food production. This book provides clarity on how this inhibitive effect of the interpretation and application of EU law may be overcome by focusing on a selection of case studies, i.e. EU consumer law, EU fundamental rights law, and EU competition law. Insights from cognitive research and behavioural economics are applied in this book to show how to overcome the inhibitive effect on sustainable consumption caused by the noncritical application of neoclassical law and economics in the interpretation and application of EU law. These insights illustrate how actual consumer behaviour sometimes systematically and predictably deviates from the normative behaviour expected from the homo economics.[31] As such, cognitive science and behavioural economics can provide insights into the "real" effect of an interpretation and application of EU law on consumer behaviour and sustainable consumption instead of an assumed effect based on normative assumptions derived from the rational actor model. Moreover, cognitive science, behavioural economics, and moral philosophy may provide insights into the effect of an interpretation and application of EU law on sustainable consumption that goes beyond material welfare by also considering happiness and capabilities that are considered important in human life.[32]

1.1 Aims and Research Questions

The theoretical aim of this book is to provide a firm understanding of how EU law can be interpreted and applied to sustainability labels. The societal aim is to align the interpretation and application of EU law with its priority to achieve sustainable consumption. To achieve these aims this book answered the following main research question:

To what extent does the interpretation and application of EU law to sustainability labels inhibit sustainable consumption?

---

[30] See e.g. the preamble to the Treaty on European Union (TEU) stating the determination "to promote economic and social progress for their peoples, taking into account the principle of sustainable development"; Article 3 TEU states that the Union's "aim is to promote (...) well-being of its peoples, which can arguably be equated with sustainable development or at least is an important component of it (see e.g. Stiglitz et al. 2009); the preamble to the Charter on Fundamental Rights (CFR) states that the EU "seeks to promote balanced and sustainable development. The EU has mainstreamed sustainable development policies through the EU Sustainable Development Strategy, the Environmental Technologies Action Plan, and the Europe 2020 Strategy, for which sustainable consumption is a priority.

[31] Tversky and Kahneman (1974), p. 1124.

[32] Adler and Posner (2008), p. 253.

To answer this main research question, an overview of different types of sustainability labels, and their capacity to promote sustainable consumption, was first provided by answering the first sub-question:

1. What is the regulatory structure underlying sustainability labels?
An overview of the types of sustainability labels gives further insight into the extent to which such labels may fall within the scope of EU law. As such, the second sub-question was:

2. When do sustainability labels fall within the scope of EU law?
Once it was clear when sustainability labels fall within the scope of EU law, the applicable legal framework to sustainability labels was assessed. Whether EU law inhibits sustainable consumption depends on its interpretation and application. The third question was, therefore:

3. To what extent does EU law leave room for interpreting and applying legal provisions to sustainability labels at the national and EU level?
Once the extent to which EU law leaves room for interpreting and applying legal provisions to sustainability labels was clear, clarity was needed as to how to ensure that legal interpretation and application does not inhibit the potential of sustainability labels to promote sustainable consumption without a legal justification. To provide such clarity, a critical examination was conducted as to whether and to what extent findings from other scientific disciplines could inform the doctrinal analysis of the interpretation and application of EU law to sustainability labels. The fourth question was, therefore:

4. Whether and to what extent could findings from other scientific disciplines inform the doctrinal analysis of the interpretation and application of EU law to sustainability labels?
The answer to these questions gives more clarity as to the actual interaction between the interpretation and application of EU law and sustainability labels. Such a clearer understanding contributes to the alignment of the objective of sustainable consumption with the application and interpretation of EU law.

1.2 Methodology

This book is essentially a doctrinal analysis to reveal the extent to which the interpretation and application of EU law interact with sustainability labels.[33] To conduct this doctrinal analysis, it is necessary to first identify the applicable legal rules. To identify the applicable rules within EU law primary and secondary sources were used. The primary sources include authoritative sources such as existing rules, principles, and precedents. The secondary sources included government documents,

[33] See on doctrinal research as a legal method at Hutchinson (2014), p. 584.

EU and international guidelines, newspapers, and scholarly publications. Based on these primary and secondary sources it was examined to what extent EU law interacts with sustainability labels. Critical legal analysis provides the theoretical framework for this book. Critical legal analysis is geared towards unfolding the policy choices behind the law.[34] In doing so, the book employs a variety of methods.

Legal frameworks are framed in specific terms or concepts.[35] To analyse the extent to which EU law could be interpreted and applied to sustainability labels, it is necessary to interpret the terms and concepts contained within the legal frameworks analysed. In this book, the methods of textual interpretation, contextual interpretation and teleological interpretation were used. These methods of interpretation are recognized by national legal orders[36] and public international law, most notably by the 1969 Vienna Convention on the Law of Treaties (VCLT).[37] As much has already been written on these methods and their use in legal interpretation,[38] the next few paragraphs will provide a brief overview of how these interpretation methods were used for this book. When necessary, resource was had to the *travaux préparatoires* as a supplementary means of interpretation.[39]

The first method of interpretation that was used was that of textual interpretation.[40] Textual interpretation may be defined as the action of explaining what a normative text conveys by looking at the usual meaning of the words contained therein.[41] General, specialized and/or technical dictionaries were examined to conduct this textual interpretation.[42] However, as emphasized by the WTO Appellate

---

[34] Hunt (1986), p. 1; Nourse and Shaffer (2009), n 11.

[35] Michaels (2012), p. 110.

[36] Bengoetxea et al. (2001), p. 48.

[37] See Articles 31-2 Vienna Convention on the Law of Treaties. The VCLT states that, in accordance with the general rule of interpretation, a treaty must be interpreted in good faith in accordance with the ordinary meaning to be given to the terms of the treaty (i.e. textual interpretation) in their context (i.e. contextual interpretation) and in the light of its object and purpose (i.e. teleological interpretation). When interpretation according to the general rule of interpretation leaves the meaning ambiguous the VCLT states that, in accordance with supplementary means of interpretation, resource may be had to the preparatory work of the treaty.

[38] See e.g. Van Damme (2010) and Lenaerts and Gutiérrez-Fons (2013–4).

[39] See for the increasing importance of the travaux préparatoires in EU law: Lenaerts and Gutiérrez-Fons (2013–4), n 38; and Case 370/12 *Pringle*, ECLI:EU:C:2012:756. See for a discussion of the use of the negotiating history by the WTO Appellate Body. Van Damme (2010), n 38.

[40] This is in line with Article 31(1) VCLT where the actual terms of the text must be the starting point of analysis. See Van Damme (2010), n 38. Furthermore, in accordance with settled ECJ case law, where the wording of an EU law provision is clear and precise, its contextual or teleological interpretation may not call into question the literal meaning of the provision, as this would run counter to the principle of legal certainty and the principle of inter-institutional balance enshrined in Article 13(3) TEU. See, e.g. Case 48/07 *Les Vergers du Vieux Tauves*, ECLI:EU:C:2008:758; Case 263/06 *Carboni e Derivati*, ECLI:EU:C:2008:128; Case 220/03 *ECB v Germany*, ECLI:EU:C:2005:748.

[41] Lenaerts and Gutiérrez-Fons (2013–4), n 38.

[42] Generally relied upon by the WTO Appellate Body. See Van Damme (2010), n 38, p. 623.

Body, 'dictionaries alone are not necessarily capable of resolving complex questions of interpretation, as they typically aim to catalogue all meanings of words – be those meanings common or rare, universal or specialized'.[43] The ordinary meaning comprises dictionary definitions and their possible uses.[44] Therefore, the meaning of a term in relation to the circumstances in which the treaty was made, and in which the language was used, was examined.[45] When treaties were officially translated in Dutch, English, French, German, and/or Spanish, these translations were examined to assess whether the translations further clarified the ordinary meaning of the assessed term.[46]

When the term was not clear and precise, or in the case of the EU, when linguistic divergences exist, textual interpretation did not suffice as a method of interpretation.[47] When this was the case, the method of contextual interpretation was used. The method of contextual interpretation is based on the idea that terms must be interpreted in a way that guarantees consistency between the term and the treaty of which it is part.[48] To ensure such consistency cross-referencing was used as an interpretive technique.[49] This technique is an application of Fitzmaurice's principle

[43]Case US-Gambling *United States-Measures Affecting Cross-Border Supply of Gambling and Betting Services (US-Gambling)*, WT/DS285/AB/R, at. 164.

[44]Van Damme (2010), n 38, p. 625.

[45]See also McNair (1961), p. 366, where he makes a distinction between the absolute and relative meaning of a term. See also Case Chicken Cuts *European Communities-Customs Classification of Frozen Boneless Chicken Cuts (EC-Chicken Cuts)*, WT/DS269/AB/R, WT/DS286/AB/R, at 175, were the WTO Appellate Body held that dictionaries were not always dispositive in finding the ordinary meaning and that 'the ordinary meaning of a treaty term must be ascertained according to the particular circumstances of each case', and that the ordinary meaning 'must be seen in the light of the intention of the parties as expressed in the words used by them against the light of the surrounding circumstances'.

[46]This is especially important in assessing EU law due to the principle of linguistic equality. See Article 55 TEU in conjunction with Article 342 TFEU, and Case 296/95 *The Queen v Commissioners of Customs and Excise, ex parte EMU Tabac and Others*, ECLI:EU:C:1998:152 at 36; Case 257/00 *Givane and others*, ECLI:EU:C:2003:8 at 36; Case 152/01 *Kyocera*, ECLI:EU:C:2003:623 at 32. The WTO Appellate Body also sometimes resorted to assessing the treaty terms in different languages. See e.g. Case EC-Asbestos *European Communities-Measures Affecting Asbestos and Asbestos-Containing Products (EC-Asbestos)*, T/DS135/AB/R, at 89, 91-2 attempting to define 'like products' assessing the French and Spanish versions of the treaty language.

[47]Lenaerts and Gutiérrez-Fons (2013–4), n 38; Case 341/01 *Plato Plastik Robert Frank*, ECLI:EU:C:2004:254 at 64; Case 236/97 *Skatteministeriet v Aktieselskabet Forsikringsselskabet Codan*, ECLI:EU:C:1998:617 at 28; Case 449/93 *Rockfon A/S tegen Specialarbejderforbundet i Danmark*, ECLI:EU:C:1995:420 at 28; Case 30/77 *Regina v Bouchereau*, ECLI:EU:C:1977:172 at 64.

[48]See e.g. Article 7 TFEU("The Union shall ensure consistency between its policies and activities, taking all of its objectives into account and in accordance with the principle of conferral of powers."); Lenaerts (2007).

[49]This technique has been used by the WTO Appellate Body as well as by the ECJ on several occasions. See for the WTO e.g. Case US-Wheat Gluten *United States-Definitive Safeguard Measures on Imports of Wheat Gluten from the European Communities (US-Wheat Gluten)*, WT/DS166/AB/R, at 72–79; see also Case US-Lamb *United States-Safeguard Measures on*

of integration that states that treaties, particular parts, chapters or sections are to be interpreted as a whole with reference to their declared or apparent objects, purposes and principles.[50]

When contextual interpretation did not suffice to interpret a term, an effective interpretation method or an evolutionary interpretation was used. The effective interpretation method is based on the principle of effectiveness.[51] After examining the normative context in with the treaty term is placed by using contextual interpretation, effective interpretation can be used to ensure that the term is interpreted as to best preserve the effectiveness of the treaty in light of its object and purpose.[52] Effective interpretation follows after contextual interpretation where the normative context in which the term is placed is examined. The principle of effectiveness can also be instrumental in justifying an evolutionary interpretation of the treaty. An evolutionary interpretation was used where the societal context required an interpretation adaptive to changes over time.[53] In the context of this book, when a term pursued more than one mutually contradictory objective, it was examined whether there existed a hierarchy between the different objectives in the respective legal order.

Imports of Fresh, Chilled or Frozen Lamb Meat from New Zealand (US-Lamb), WT/DS177/AB/R, WT/DS178/AB/R at 162–181; Case US-Line Pipe *United States-Definitive Safeguard Measures on Imports of Circular Welded Carbon Quality Line Pipe from Korea (US-Line Pipe)*, WT/DS202/AB/R, at 209–211. Compare with Case Concerning the Auditing of Account between the Kingdom of the Netherlands and the French Republic pursuant to the Additional Protocol of 25 September 1991 to the Convention of 3 December 1976 on the Protection of the Rhine against Pollution by Chlorides, PCA Award Series (2008) 107, at. 91. For the use of cross-referencing by the ECJ see e.g. Case 465/07 *Elgafaji*, ECLI:EU:C:2009:94; see also Lenaerts (2010), p. 255 who provides a commentary on the merits of that case.

[50] Fritzmaurice (1957), pp. 9 and 211. See also, e.g., Case International Status of South-West Africa, Dissenting Opinion of Judge de Visscher *International Status of South-West Africa, Dissenting Opinion of Judge de Visscher*, ICJ Reports 1950 at 187; Case Case Concerning Right of Nationals of the United States of America in Morocco (France v. United States of America) *Case Concerning Right of Nationals of the United States of America in Morocco (France v. United States of America)*, ICJ Reports 1952, at 209; Case Ambatielos Case (Greece v. United Kingdom) (Merits: Obligation to Arbitrate), Joint Dissenting Opinion of Judges McNair, Basdevant, Klaestad, and Read *Ambatielos Case (Greece v. United Kingdom) (Merits: Obligation to Arbitrate), Joint Dissenting Opinion of Judges McNair, Basdevant, Klaestad, and Read*, ICJ Reports 1953 at 30; Emer De Vattel, *The Law of Nations, or, Principles of Law of Nature Applied to the Conduct and Affairs of Nations and Sovereigns* (4th edn, 1811), pp. 255–256.

[51] See for the principle of effectiveness in international law: Van Damme (2010), n 38, p. 625. See on functional interpretation in ECJ case-law Case 439/08 *VEBIC*, ECLI:EU:C:2010:739 at 64; Case 188/10 *Melki and Abdeli*, ECLI:EU:C:2010:363 at 145; Case 409/06 *Winner Wetten*, ECLI:EU:C:2010:503 at 56.

[52] Lenaerts and Gutiérrez-Fons (2013–4), n 38.

[53] See on evolutionary interpretation Bernhardt (1999), p. 11; Helmersen (2013), p. 127.

The doctrinal analysis of this book went beyond a mere interpretation of the current law.[54] A critical reflection of the current laws was conducted, suggesting ways the philosophy and processes underlying the laws could be improved to promote sustainable development. To provide context for such a critical reflection, the assessed legal frameworks were, if relevant, compared with the USA legal system. To compare the EU legal system and the USA legal system a functional method was used.[55] By using a functional comparative method the purpose and function of the specific rules were identified, to interpret and evaluate how well the rule serves its purpose or function. As such, the comparison provided more clarity as to the purposes of the legal rules and other ways that these purposes could be accomplished.

To provide further context for critical reflection, the legal analysis was placed within the larger framework of the social sciences. Such an interdisciplinary analysis enabled a more scientifically grounded analysis of the effect of the interpretation of legal terms and concepts on sustainable development. For example, when a court mandates disclaimers on food labels to clarify otherwise misleading sustainability claims, insights from cognitive science and behavioural economics may clarify whether such disclaimers would have a positive or negative effect on consumers to purchase sustainably. Another example is that when rules aim to promote consumer welfare, without clarifying how such consumer welfare could be estimated, insights from economics, psychology, and philosophy may be helpful to estimate consumer welfare in a way that ensures, or at least not deters, sustainable development.

As sustainability labels are international it is important to see beyond a single legal system and consider EU law. To conduct the doctrinal analysis of the EU legal provisions that apply to sustainability labels, a case study analysis was selected which provided an extensive examination of how EU law and sustainability labels interact. To provide context for critical reflection within the case study analysis, insights from comparative legal methods and methods from the social sciences were used. As such, the case study analysis serves as a showcase of how insights from the social sciences, most notably cognitive research and behavioural economics, could be integrated in doctrinal analysis to align the objective of sustainable development with the interpretation and application of EU law to sustainability labels.

The case study analysis provides a description and examination of three legal frameworks, examining the inhibitive effect on sustainable consumption of different interpretations and applications legal frameworks to sustainability labels. The selected case studies were the legal frameworks that may be of particular relevance to sustainability labels: EU consumer law, EU fundamental rights law, and EU competition law. These legal frameworks were selected as their current interpretation and application illustrate the possible inhibitive effect of EU law on sustainability

[54] See Hutchinson (2015), pp. 130, 132. who clarifies that 'good' quality doctrinal research should go "well beyond description, analysis, and critique, and invariably suggests ways the law could be amended or the philosophy, processes or administration of the law could be improved".

[55] See on the functional method, e.g. Michaels (2012), n 35; and Gordley (2012), p. 113.

consumption. The research concentrates on sustainability labels on food products because food production bears the largest environmental costs compared to other sectors,[56] and food consumption may affect most categories often associated with sustainable development (e.g. organic, fair trade, human health, animal welfare).

1.3 Structure of the Book

This book starts with an overview of the regulatory structure of sustainability labels in Chap. 2. To provide such an overview the regulatory structure underlying 65 sustainability food labels in the Dutch market were examined. To identify the presence of a regulatory structure underlying sustainability labels three classes of criteria were used to benchmark sustainability labels on the Dutch market: (1) the aim the standard is trying to achieve; (2) the actors involved in standard-setting; and (3) the type of certification scheme used.

In Chap. 3, the extent to which the interpretation and application of EU consumer law to sustainability labels inhibit sustainable consumption was investigated. It was examined whether and how insights from behavioural sciences could inform the legal interpretation and application of the normative EU "average consumer" benchmark under the Food Information Regulation (FIR)[57] when confronted with purely visual information as used by green pictograms. To conduct such an analysis, the relevant legal norms were, first, located by using primary and secondary sources. To interpret the legal norms as to analyse whether it prohibits "misleading impressions of sustainability", the legal analysis relied on the leading cases of the ECJ on misleading information, as well as the relevant EU legal texts, policy documents and scholarly papers. To analyse whether consumers in a real-life setting could be considered misled by a sustainability label, an experiment was conducted. This empirical analysis relied on questionnaires. These questionnaires were used to collect information from consumers as to whether a specific food label, that is present on the Dutch market, can be considered to give a misleading impression of sustainability.

In Chap. 4, the extent to which the interpretation and application of EU fundamental rights law to sustainability labels inhibit sustainable consumption was explored. It was examined whether the freedom of expression as enshrined in Article 11 of the CFR could be interpreted to limit the EU's freedom to regulate sustainability labels. It was explored to what extent the freedom of expression protects food businesses against government intervention with communications on food labels. Practice in the USA shows that governmental regulations of commercial

[56] KPMG, *Expect the Unexpected: Building Business Value in a Changing World* (Report: KPMG International, 2012).

[57] Regulation (EU) 1169/2011 on the provision of food information to consumers (EU Food Information Regulation) [2011] OJ L 304.

communication could be considered inconsistent with the freedom of expression of businesses.[58] Although such practice is not yet apparent in the EU, recent trends show that governmental regulations of commercial communications could be considered inconsistent with the freedom of expression of businesses.[59] To assess whether government interventions with communications on food labels protect food businesses against government intervention a functional comparison between the two legal systems was conducted.[60] The analysis was placed in the larger framework of social sciences by using insights from economic and social psychology as to how consumers process information on food products.

In Chap. 5, the extent to which the interpretation and application of EU competition law, more specifically Article 101 TFEU, to sustainability labels inhibit sustainable consumption was examined. This analysis required, first, an assessment of how Article 101 TFEU is currently applied to private agreements based on primary sources, such as EU legal texts, case-law by the ECJ, and notices by the European Commission, and secondary sources, such as scholarly papers. Secondly, this analysis required an assessment as to the underlying assumptions in the interpretation and application of Article 101 TFEU to private sustainability initiatives. These assumptions should first be determined based on primary sources and secondary sources. To conduct a critical reflection of the current legal practice and to provide effective tools to align sustainable development with the interpretation and application of Article 101 TFEU to sustainability labels, the legal analysis was placed in the larger framework of the social sciences by including insights from welfare economics, economic and social psychology, and moral philosophy.

Chapter 6 contains the conclusions and general discussions of this book. It was concluded that, when properly applied, the approaches proposed in this book are better equipped to align the objective of sustainable development with the interpretation and application of EU law to sustainability labels, than solely using approaches derived from neoclassical law and economics. The contributions this book makes to legal and policy practice, and legal research and methodology in the application and interpretation of EU law were briefly discussed.

The overall structure of the book can be found in Table 1.1.

---

[58] See for example, Case Amestoy *Amestoy*, 92 F.3d 67 (2nd Cir. 1996); Case Pearson I *Pearson I*, 164 F.3d 650 (D.C. Cir. 1999), rehearing den., 172 F.3d 72 (D.C. Cir. 1999); Case Whitaker v Thompson *Whitaker v Thompson*, 48 F. Supp. 2d 1 (D.D.C. 2002); Case Boggs *Boggs*, 622 F.3d 628 (6th Cir. 2010).

[59] In Recital 44 of Regulation (EU) No 609/2013 of the European Parliament and of the Council of 12 June 2013 on food intended for infants and young children, food for special medical purposes, and total diet replacement for weight control it is stated that "[t]his Regulation does not affect the obligation to respect fundamental rights and fundamental legal principles, including the freedom of expression, as enshrined in Art. 11, in conjunction with Art. 52, of the Charter of Fundamental Rights of the European Union, and in other relevant provisions." Furthermore, Case 544/10 *Weintor*, ECLI:EU:C:2012:526 was the first, and so far only, ECJ-case concerning a fundamental rights challenge against a food labelling law.

[60] See on the functional method, e.g. Michaels (2012), n 35; and Gordley (2012), n 55, p. 113.

Table 1.1 Structure of the book

Chapter	Title	Answer to research question
1	Introduction	
2	Regulatory Structure Underlying Sustainability Labels	1
3	Sustainability Labels in the Shadow of EU Consumer Law	2, 3, 4
4	Sustainability Labels in the Shadow of EU Fundamental Rights Law	2, 3, 4
5	Sustainability Labels in the Shadow of EU Competition Law	2, 3, 4
6	Conclusion and Discussion	

References

Adler M, Posner E (2008) Happiness research and cost-benefit analysis. J Leg Stud 37:253

Akerlof GA (1970) The market for "lemons": quality uncertainty and the market mechanism. Q J Econ 84:488

Becher S, Feldman Y (2016) Manipulating, fast and slow: the law of non-verbal market manipulations. Cardozo Law Rev 38

Bengoetxea J, Soriano LM, McCormick N (2001) Integration and integrity in the legal reasoning of the European Court of Justice. In: de Búrca G, Weiler JHH (eds) The European Court of Justice. Oxford University Press, p 48

Bernhardt R (1999) Evolutive treaty interpretation, especially of the European Convention on Human Rights. German Yearb Int Law 42:11

Cafaggi F, Iamiceli P (2013) Supply chains, contractual governance and certification regimes. Eur J Law Econ 37:131

Darby M, Karni E (1973) Free competition and the optimal amount of fraud. J Law Econ 16:67

Faure M, Luth H (2011) Behavioural economics in unfair contract terms – cautions and considerations. J Consum Policy 34

Franck J-U, Purnhagen K (2014) Homo Economicus, behavioural sciences, and economic regulation: on the concept of man in internal market regulation and its normative basis. In: Mathis K (ed) Law and economics in Europe: foundations and applications. Springer, Netherlands

Fritzmaurice G (1957) The law and procedure of the International Court of Justice 1951–4: treaty interpretation and other treaty points. Br Yearb Int Law 33:9 and 211

Gordley J (2012) The functional method. In: Monateri PG (ed) Methods of comparative law. Edward Elgar Publishing, p 113

Green KC, Scott Armstrong J (2012) Evidence on the effects of mandatory disclaimers in advertising. J Public Policy Market 31:293

Grunert K, Hieke S, Wills J (2014) Sustainability labels on food products: consumer motivation, understanding and use. Food Policy 44:177

Hahnel U et al (2015) The power of putting a label on it: green labels weigh heavier than contradicting product information for consumers' purchase decisions and post-purchase behavior. Front Psychol 6:1392

Heinemann A (2015) In: Mathis K (ed) A "more realistic approach" to competition law. Springer

Helmersen ST (2013) Evolutive treaty interpretation: legality, semantics and distinctions. Eur J Leg Stud 6:127

Horne R (2009) Limits to labels: the role of eco-labels in the assessment of product sustainability and routes to sustainable consumption. Int J Consum Stud 33:175

Howells G, Weatherill S (2005) Consumer protection law. Ashgate Publishing Group, Aldershot

Hunt A (1986) The theory of critical legal studies. Oxford J Leg Stud 6:1

Hutchinson T (2014) Valé Bunny Watson? Law librarians, law libraries and legal research in the post-internet era. Law Libr J 106:579
Hutchinson T (2015) The doctrinal method: incorporating interdisciplinary methods in reforming the law. Erasmus Law Rev 130:132
Incardona R, Poncibò C (2007) The average consumer, the unfair commercial practices directive, and the cognitive revolution. J Consum Policy 30:21
Lenaerts K (2007) The rule of law and the coherence of the judicial system of the European Union. Common Mark Law Rev 44
Lenaerts K (2010) The contribution of the European Court of Justice to the area offreedom, security and justice. Int Comp Law Q 59:255
Lenaerts K, Gutiérrez-Fons JA (2013–4) To say what the law of the EU is: methods of interpretation and the European Court of Justice. Columbia J Eur Law 20
McNair A (1961) The law of treaties, 2nd edn. Oxford University Press, p 366
Michaels R (2012) Method of comparative law. In: Reimann M, Zimmermann R (eds) The Oxford handbook of comparative law. Oxford University Press, p 110
Micklitz HW (2008) Some reflections on Cassis de Dijon and the control of unfair contract terms in consumer contracts. In: Collins H (ed) Standard contract terms in Europe: a basis for and a challenge to European contract law. Kluwer Law International
Monti G, Mulder J (2017) Escaping the clutches of EU competition law: pathways to assess private sustainability initiatives. Eur Law Rev 42:641
Nelson P (1970) Information and consumer behavior. J Polit Econ 78:311
Nourse V, Shaffer G (2009) Varieties of new legal realism: can a new world order prompt a new legal theory. Cornell Law Rev 95:61
Posner R (1997) Rational choice, behavioral economics, and the law. Stanford Law Rev 50:1551
Purnhagen K (2017) More reality in the CJEU's interpretation of the average consumer benchmark – also more behavioural science in unfair commercial practices? Eur J Risk Regul 8:437
Schebesta H, Purnhagen K (2016) The behaviour of the average consumer: a little less normativity and a little more reality in the court's case law? Reflections on Teekanne. Eur Law Rev 41:590
Sibony A-L, Helleringer G (2015) EU consumer protection and behavioural sciences: revolution or reform? In: Alemanno A, Sibony A-L (eds) About nudge and the law. Hart
Simon H (1955) Behavioural model of rational choice. Q J Econ 69:99
Spadoni R, Lombardi P, Canavari M (2014) Private food standard certification: analysis of the BRC standard in Italian agri-food. Br Food J 116:142
Stiglitz J, Sen A, Fitoussi J-P (2009) Report by the Commission on the measurement of economic performance and social progress. Paris
Trzaskowski J (2011) Behavioural economics, neuroscience, and the unfair commercial practises directive. J Consum Policy 34:377
Tversky A, Kahneman D (1974) Judgment under uncertainty: heuristics and biases. Science 185: 1124
Van Dam Y, Van Trijp H (2011) Cognitive and motivational structure of sustainability. J Econ Psychol 32:726
Van Damme I (2010) Treaty interpretation by the WTO appellate body. Eur J Int Law 21
Zerbe N (2004) Feeding the famine? American food aid and the GMO debate in Southern Africa. Food Policy 29:593

Chapter 2
Regulatory Structure Underlying Sustainability Labels

2.1 Introduction

Sustainability labels have proven very popular to provide sustainability-related information to consumers. Since the 1980s there has been a rise of sustainability labels on the consumer market.[1] For example, in 2014 around 170 different sustainability labels were being used in The Netherlands.[2] Despite the rise of sustainability labels since the 1980s the market share of sustainable food remains low.[3] Different reasons for this slow increase in market share for sustainable food may include the wide diversity of sustainability labels as shown by research on the credibility of eco-labels,[4] limited consumer understanding of specific sustainability labels,[5] and/or failure of such sustainability labels to trigger consumer motivation to purchase sustainably.[6]

Intervention in the plethora of sustainability labels on the market may be necessary to increase their capacity to promote sustainable consumption. To determine if,

---

This book chapter is based on Eva van der Zee, 'Investigating the Regulatory Structure of Voluntary Sustainability Standards: Foundations for Intervention Strategies to Increase Consumer Confidence. In: Regulating and Managing Food Safety in the EU' (2018), Springer, pp. 39–55.

---

[1] Grunert et al. (2014); Gandenberger et al. (2011), p. 107; Koos (2011), p. 127; Raynolds et al. (2007), p. 147; Wilkinson (2007), p. 219.

[2] ANP. 2014, April 3. Ploumen wil 'orde' in aantal keurmerken (English translation: Ploumen wants 'order' in the amount of certification marks). Volkskrant. Retrieved from www.volkskrant.nl/binnenland/ploumen-wil-orde-in-aantal-keurmerken~a3627943/.

[3] See e.g. Dutch Ministry of Economic Affairs 2014. Consumentenbestedingen aan Duurzaam Gelabelde Producten. Monitor Duurzaam Voedsel 2013. Although it is steadily increasing in the last few years from a market share of 5.5% in 2012 to a market share of 14% in 2019. See Katja Logatcheva, 'Montior Duurzaam Voedsel', https://edepot.wur.nl/532565.

[4] Fischer et al. (2005); Teisl et al. (2002), p. 44.

[5] Horne (2009); Grunert et al. (2014), n 1.

[6] Van Dam and De Jonge (2015), p. 19.

© The Author(s), under exclusive license to Springer Nature Switzerland AG 2022

E. van der Zee, *Sustainability Labels in the Shadow of the Law*, Studies in European Economic Law and Regulation 22, https://doi.org/10.1007/978-3-030-95802-2_2

when and where public or private institutions should intervene an overview of the current sustainability labels landscape is needed. This chapter provides such an overview and identifies the knowledge gaps that are further explored in Chaps. 3, 4 and 5.

2.2 Theoretical Framework

Business-to-consumer information labelling provides an increasingly important regulatory tool, promoting informed consumer choice by communicating sustainability attributes of products or production processes through logos, symbols, labels, and texts. The extent to which a labelling scheme provides relevant sustainability-related information depends on how sustainability is conceived. Numerous definitions of sustainability, or sustainable development, have appeared in international declarations, conventions, court judgments, and scholarly writings.[7] Although definitions differ, many definitions emphasise that the needs of the present must be met, while the ability of future generations to meet their own needs may not be compromised.[8] This is also known as inter- and intra-generational equity. Many definitions further specify economic, social, environmental and sometimes cultural aspects as important dimensions of sustainable development.[9] From these more abstract aspects of sustainability, more concrete sub-categories of sustainability can be determined, e.g. biodiversity, human rights and labour conditions. These categories may address several issues, e.g. climate change, child labour and wage levels (see also Table 2.1).

As many sustainability attributes of products can be expensive or impossible to determine even after purchase, consumers may question the credibility of a sustainability label.[10] Certification has as a function to assure consumers that the properties which the sustainability label represents exist. Certification can be defined as the assessment and approval by a party of a standard.[11]

Certification may involve three possible parties: the certification applicant, the certification body, and the accreditation body. The certification applicant is the manufacturer or distributor that requests certification of its product or

[7] Johnston et al. (2007), p. 60; Lawn (2010).

[8] World Commission on Environment and Development, *Our Common Future* (1987).

[9] Boström (2012), p. 3; Connelly (2007), p. 259; De Boer et al. (2004); Norgaard (1988), p. 606; Redclift (1992), p. 395; Organisation for Economic Co-operation and Development (OECD) 2008. *Promoting Sustainable Consumption*. Good Practices in OECD countries; International Institute for Environment and Development (IIED) 1997. Changing Consumption and Production Patterns: Unlocking Trade Opportunities. *UN Department of Policy Coordination and Sustainable Development*, http://pubs.iied.org/pdfs/8851IIED.pdf.

[10] See Chap. 1 on credence and Potemkin attributes of sustainability.

[11] Meuwissen et al. (2003).

Table 2.1 Sustainability ideal, aspects, categories, and issues potentially relevant in the context of sustainability labels

Ideal	Aspects	Categories	Issues
Inter- and intra-generational equity	Environmental aspects	Environmental impacts Biodiversity	Climate change Environment and health Natural resources and waste Biodiversity in general One or more specific ecosystems Specific species
	Social aspects	Human health Human rights Labour conditions Animal welfare	Food safety Forced labour Rights of indigenous people Freedom of association/bargaining Child labour Wage levels Occupational health and safety standards Working hours Free-range
	Economic aspects	Trade relations	Guaranteed price Long-term contracts Advanced payments and credit facilities Technical assistance Community support
	Cultural aspects	Geographical indications Traditional specialities Religion	Designation of origin Geographical indication Traditional character Halal Kosher

Based on De Boer et al. (2004), n 9

production-process. The certification body is the actor that issues the certification. The accreditation body is a formal third party that complies with the general requirements for accreditation bodies specified by the International Standard ISO/IEC 17011:2004 set by the International Standard Organization (ISO). In general, accreditation bodies monitor certification bodies on compliance with the International Social and Environmental Accreditation and Labelling (ISEAL) alliance assurance code and/or the ISO 17065 (former ISO 65 or EN 45011).

Basically, certification schemes include public (state-run) schemes and private certification schemes.[12] Public certification schemes are assessed and approved by public bodies (Fig. 2.1). Three EU quality schemes are prime examples of such

[12] Jahn et al. (2005), p. 53.

Fig. 2.1 Public certification

Fig. 2.2 First-party certification

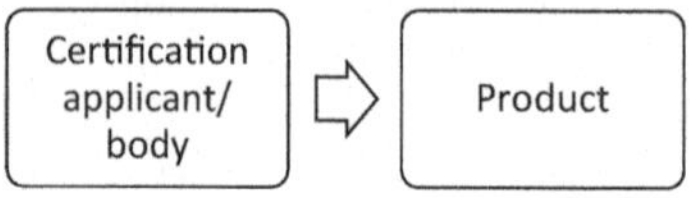

Fig. 2.3 Second-party certification

public certification schemes, i.e. PDO (protected designation of origin), PGI (protected geographical indication) and TSG (traditional speciality guaranteed). The European Commission examines whether a food product can be registered as PDO, PGI, or TSG. Public bodies control products registered as PDO, PGI, or TSG on compliance with EU legislation.[13]

Private certification schemes can take different forms: first-party certification, second-party certification, and third-party certification.[14] First-party certification is when the manufacturer or distributor of the food product itself sets the standards and certifies compliance of the product with the set standard (Fig. 2.2). Many examples of first-party certification can be found with respect to seafood products, e.g. logos like "dolphin-friendly tuna" and "fished by pole & line". In general, these standards are not externally checked. Other examples include standards such as AH Puur & Eerlijk,[15] Fairglobe, or Bio+. These sustainability labels can be considered umbrella standards as the standards umbrellas for other sustainability labels: the umbrella standard is first-party certified but requires that the product is certified through specific third-party certification systems.

Second-party certification is conducted by an organisation other than the certification applicant (Fig. 2.3). These organisations are legally separate from the certification applicant. An example of second-party certification is the V-label. The V-label was launched by the European Vegetarian Union (EVU): an umbrella organisation for most vegetarian groups in Europe. The mark is checked by the

[13] In The Netherlands the control bodies are: NVWA (government agency), COKZ and KCB (both government bodies).

[14] Marx (2011).

[15] This sustainability label no longer exists after strong societal critiques. See Volkskrant, "AH draait huismerk 'Puur en Eerlijk' de nek om", 8 January 2015. However, the label has been praised by academics for being more effective in motivating occasional sustainable consumers to increase their sustainable purchase behaviour. van Dam and van Trijp (2011), p. 726.

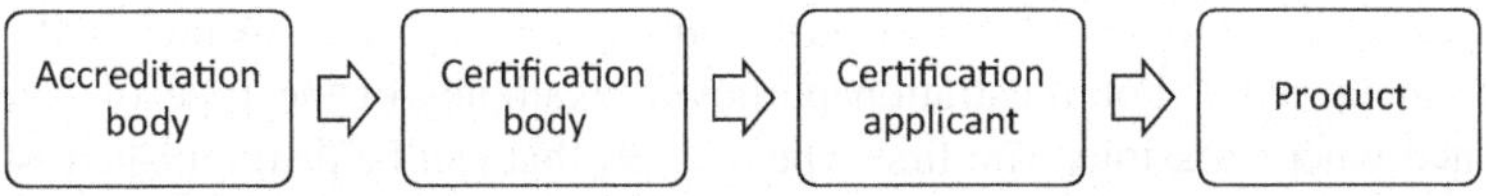

Fig. 2.4 Third-party certification

different national vegetarian organizations per country, i.e. Nederlandse Vegetariërsbond (Dutch Vegetarian Society).[16]

Third-party certification is conducted by a certification body that is legally separate and independent from the organisation that asks for certification (Fig. 2.4). The certification body needs to be accredited by an accreditation body to ensure its competence, impartiality and performance capability. An example of third-party certification is the EU organic standard. Every Member State of the EU has at least one private certification body responsible for the certification. This certification body has to be accredited.[17] Due to the independence of the certification body, third-party certification is gaining popularity on all levels of the agri-food chain.[18]

Besides the credibility of the certification scheme, the credibility of a standard may also depend on the credibility of the actors involved in setting the standard.[19] The type of actors involved in certification are important as especially lack of clarity over the source of an eco-label is a key factor inducing consumer uncertainty.[20] Thus, to prevent market failure, not only the type of certification but also the types of actors involved need to be considered credible.

Three types of actors can be involved in certification: public actors, NGOs and firms.[21] Public actors encompass developing and developed states, as well as governmental agencies. NGOs include all private actors that are not targeted by the sustainability label, e.g. NGO advocacy groups, labour unions, non-profit organisations student groups, and other civil society organisations, as well as socially responsible investors. Firms include those private actors that *are* targeted by the sustainability label, e.g. multinationals, small firms, agricultural enterprises, and small-scale farmers. Consumer research suggests that consumers may have lower confidence in company-owned standards.[22] They seem to have higher confidence in a sustainability label that is perceived to be independent.[23]

[16] http://www.v-label.info.

[17] Article 5 (2) (c) of Regulation (EC) 882/2004 on official controls performed to ensure the verification of compliance with feed and food law, animal health and animal welfare rules [2004] OJ L 165.

[18] Auriol and Schilizzi (2003).

[19] Albersmeier et al. (2010), p. 69; Moussa and Touzani (2008), p. 526.

[20] Harbaugh et al. (2011), p. 1512.

[21] Abbott and Snidal (2009), p. 501.

[22] Eden et al. (2008), p. 1; Sønderskov and Daugbjerg (2011), p. 507.

[23] Padel and Foster (2005), p. 606; Eden et al. (2008), n 24.

Based on the actors involved in certification, seven types of sustainability labels can be identified.[24] For illustration purposes, examples of the type of labels are presented where possible. The first type of label that can be distinguished is set by public actors only, which may include national laws and regulations, but also VSS. Examples of VSS set by public actors are the EU organic standard and the EU quality schemes for geographical indications and traditional specialities. The second type of label is set by firms only, which may include self-regulative schemes, such as dolphin-safe-tuna or fished by pole and line, and brands, such as AH Puur & Eerlijk and Bio+. The third type of label is set by NGOs only, which may include labels such as Vegan, V-keurmerk, and the Checkmark (Vinkje in Dutch). The fourth type of label is set in collaboration with public actors and firms. The UN Global Compact is an example of such a label, aiming at aligning business operations and strategies with principles in the areas of human rights, labour, environment and anti-corruption.[25] The fifth type of label is set by a collaboration between public actors and NGOs. Although there are not many examples of such standards, one example is the TCO certification scheme for office equipment.[26] The sixth type of standard is set by a collaboration between firms and NGOs, which may include the more well-known labels, such as Max Havelaar, UTZ Certified, and Rainforest Alliance. The seventh type of label is set by a collaboration between public actors, firms, and NGOs. An example outside the food domain is the Kimberley Process Certification Scheme focusing on conflict diamonds.[27]

2.3 Methodology

To identify the presence of a regulatory structure underlying sustainability labels three classes of criteria need to be examined: (1) the aim the label is trying to achieve; (2) the actors involved in standard-setting; and (3) the type of certification scheme used. These three classes of criteria were used to benchmark sustainability labels on the Dutch market. These sustainability labels were selected from the *Keurmerkenwijzer* of Milieucentraal.[28] The *Keurmerkenwijzer* presents over 170 labels, symbols, and logos that seemingly provide information about the sustainability of food, clothes, cosmetics, detergents, appliances, wood, paper, flowers and plants. These labelling schemes could be found on products in The Netherlands in 2014. Of these 170 labelling schemes, 65 are visual representations of

[24] Abbott and Snidal (2009), n 23.

[25] https://www.unglobalcompact.org.

[26] http://tcodevelopment.se.

[27] http://www.kimberleyprocess.com/.

[28] The Milieu Centraal Foundation is an independent information organisation that offers information to consumers on environmental and energy issues in daily life. www.milieucentraal.nl/over-milieu-centraal/.

sustainability on food labels.[29] For this study, these 65 visual representations were selected (Appendix 1).[30]

These sustainability labels were examined based on three topics: (1) aim; (2) actors involved in certification; and (3) types of certification. The information about these three topics was found through desk research by examining webpages of each visual representation and by consulting other websites such as www.keurmerkenwijzer.nl, www.labelinfo.be,[31] www.label-online.de,[32] and www.standardsmap.org.[33]

2.4 Results

2.4.1 Diversified Aim of Sustainability Labels

The aim of the sustainability labels was diverse (see Table 2.2). Of the four different aspects of sustainability, nine different combinations were observed. Sustainability labels mainly focused on environmental and social aspects (24), social aspects only

Table 2.2 Sustainability focus of sustainability labels

Aspects	Total	Of which social is animal welfare only
Environmental + social	24	12
Social	11	7
Environmental + social + economic	7	0
Cultural	5	N/A
No information	5	N/A
Environmental	4	N/A
Environmental + social + cultural	3	3
Social + economic	3	0
Environmental + cultural	2	N/A
Environmental + economic + cultural	1	N/A
Total	65	22

[29] Beter Leven Kenmerk had three entries in de *keurmerkenwijzer* based on the Beter Leven Kenmerk ranking system: 3 stars, 2 stars, and one star. For the purposes of this study Beter Leven Keurmerk had only one entry.

[30] The author, nor Wageningen University, specifically recommends any of these labelling schemes and/or find these labelling schemes better or worse that those that were not selected for the study.

[31] Website of Belgium consumers organisation Netwerk Bewust Verbruiken in cooperation with OIVO en Ecoconso, and support of Cel Sociale Economie, the POD DO, the FOD Leefmilieu and Leefmilieu Brussel. The website contains information about sustainability labels that are also present on Dutch food products.

[32] Website of independent German consumer organisation Verbraucher Initiative.

[33] Website initiated by the International Trade Centre, the joint agency of the World Trade Organization (WTO) and the United Nations Conference on Trade and Development (UNCTAD).

Table 2.3 Requirements relating to husbandry practices and housing conditions of dairy cows of EU Organic, Demeter, Duurzame Weidezuivel and Weidemelk

Sustainability label	Type of access	Length of time
EU Organic	Open air, preferably pasture	Permanent
Demeter	Open air, preferably pasture	Permanent
Duurzame Weidezuivel	Pasture	130 days a year, 8 h a day
Weidemelk	Pasture	120 days a year, 6 h a day

(11), and environmental, social and economic aspects (7). Frequently, when standards focused on social aspects, the focus was solely on the category animal welfare.

Although some sustainability labels focused on similar aspects of sustainability, this does not necessarily imply that the requirements are similar. These specific requirements are, however, often difficult to find and, therefore, difficult to compare. The specific requirements that could be found often differed only slightly. For example, Demeter and EU Organic aim at environmental aspects and animal welfare. Demeter sets, however, additional production and processing requirements, e.g. as to mutilations of poultry (like forbidding cutting, trimming, or castration,[34] at least two-thirds of the fodder offered to the animals must originate from Demeter production,[35] and milk may not be homogenised.[36] When EU Organic and Demeter are compared on husbandry practices and housing conditions of dairy cows with the husbandry practices and housing condition requirements of two other dairy standards that focus on animal welfare, Duurzame Weidezuivel (Sustainable Pasture Dairy) and Weidemelk (Pasture Milk), differences also occur (see Table 2.3).

2.4.2 *High NGO Involvement in Standard-Setting*

Sustainability labels are mainly created by NGOs (24), and to a lesser extent also by firms (14), public actors (14), or through a partnership between NGOs and firms (10). Partnerships between NGOs and firms generally took the form of NGOs having public consultations with firms before standard-setting. Public-private partnerships, i.e. public actors collaborating in standard-setting with firms, NGOs, or both, were rare in the sample. There were only three public-private partnerships and these were between public actors and NGOs: KRAV, Biogarantie, and Soil Association. These public-private partnerships were, moreover, sustainability labels regulating organic production processes that were originally founded by NGOs; later these became

[34] Article 5.4.4. Production Standards For the Use of Demeter Biodynamic® and Related Trademarks As of June 2014 (to be implemented by each member country by the 1st July 2015).

[35] Ibid., Article 5.5.1.

[36] Ibid., Article 4.1.

Table 2.4 Actors involved in standard-setting and type of certification

	Public actor	Public actor + NGO	NGO	NGO + firm	Firm
Public certification	4	0	0	0	0
Third-party	9	3	12	8	1
Second-party	1	0	10	0	0
First-party	0	0	2	1	11

official national organic standards that had to meet the requirements of the EU Regulation on Organic Agriculture[37] but upheld their stricter requirements.

2.4.3 *Certification of Standards Often Through Private, Third-Party Certification*

Most sustainability labels were certified through certification schemes. For three sustainability labels, it was unclear what type of certification scheme was used for certification. Only four sustainability labels, i.e. the Danish organic Ø-mark and the EU schemes PDO, PGI, and TSG, were certified through public certification schemes. The other sustainability labels were certified through private certification. Most sustainability labels are certified through third-party certification (33),[38] followed by first-party certification (14), and second-party certification (11).

The 33 sustainability labels that are certified through third-party certification are often set by NGOs (12), public actors (9), or a partnership between NGOs and firms (8) (see Table 2.4). Sustainability labels that are certified through second-party certification are often set by NGOs only, while VSS that are certified through first-party certification are often set by firms only.

2.5 Conclusion and Discussion

This study shows that a regulatory structure underlies sustainability labels with respect to their aim, actors involved in standard-setting, and their certification scheme used. Most sustainability labels have two main characteristics, which can be found within each group, e.g. animal welfare standards, fair trade standards, or environmental standards. First, sustainability labels often have high NGO involvement in standard-setting. Second, certification of these standards is mainly conducted by private certifiers, through third-party certification. Wittingly or

[37]Council Regulation (EC) 834/2007 on organic production and labelling of organic products (EC Regulation on Organic Agriculture) [2007] OJ L 189.

[38]Ecosocial performed two types of certification: third party certification for organic produce, and second-party certification for non-organic produce. Ecosocial is, therefore, counted twice.

unwittingly the majority of sustainability standards resemble a structure that is trusted by consumers and considered most reliable and credible according to consumer research. First, private, third-party certification is generally perceived as more reliable and credible than first- or second-party certification[39] and thus more effective in ensuring food quality.[40] Second, consumer research suggests that consumers have more confidence in NGOs than in firms or governments in standard-setting.[41] Based on these insights it may seem that the majority of sustainability labels are on the right track.

Even though the majority of sustainability labels may be on the right track, several sustainability labels may not. Current results show that almost half of the sustainability labels have no NGO involvement in the standard-setting process. Furthermore, certification of almost half of the sustainability labels is not conducted through third-party certification, but through first- or second-party certification. In other words, the majority of sustainability labels may be considered ‘good’, i.e. they are more reliable, credible,[42] effective,[43] and trusted,[44] while a minority may be considered ‘bad’, i.e. they are not considered as reliable, credible, effective, and trusted. The coexistence of a plurality of ‘good’ and ‘bad’ sustainability labels on the market may confuse consumers, as consumers cannot trust that every sustainability label is ‘good’. As this study also showed the multiplicity and diversity of sustainability labels on the market, it is very difficult for consumers to know every single sustainability label and distinguish ‘good’ from ‘bad’ labels. As such, consumer perception towards the reliability and credibility of sustainability labels, even to those labels that are reliable and credible, may be negatively affected. As a consequence, the effectiveness of sustainability labels to increase sustainable consumption may be reduced.

To increase the effectiveness of sustainability labels to increase sustainable consumption, private and/or public intervention may be necessary. An example of private intervention is the merge between Rainforest Alliance and UTZ Certified by creating a single agricultural sustainability label to simplify the certification process and increase engagement with consumers.[45] As actors are likely to commit themselves to a system of governance when they grasp that their self-interest is served by such commitment,[46] it is not unlikely that further collaborations will emerge. Governments could induce such collaborations between private standard-setters by drafting international procedural guidelines for standard-setting.[47]

[39] Golan et al. (2001), p. 117.

[40] Hatanaka et al. (2005), p. 354.

[41] Padel and Foster (2005), n 25; Eden et al. (2008) ibid, n 24.

[42] Golan et al. (2001).

[43] Hatanaka et al. (2005), n 42.

[44] Padel and Foster (2005), n 25; Eden et al. (2008), n 24.

[45] http://www.rainforest-alliance.org/article/rainforest-alliance-utz-merger.

[46] Druzin (2014), p. 1049.

[47] Van der Zee (2018).

An example of public intervention is governments setting, through command and control regulatory instruments, minimum requirements which a product or production process claiming to be sustainable should meet. For example, in the US, when energy efficiency was becoming popular, a great diversity of claims was made about the value of energy efficiency. The US government stepped in and created a third-party certified energy efficiency standard for products and buildings that meet the highest energy efficiency levels.[48] Consumers could recognise products and buildings that met this standard through the state-controlled, voluntary standard called Energy Star. Another example of governments that stepped in when a great diversity of sustainability labels appeared on the market with respect to sustainability issues is the EU initiated organic standard. The EU organic standard is a voluntary, third-party certified standard. Producers are legally obliged to place the stamp accompanying the standard on their product labels when they claim their product to be organic. The EU organic standard was set up to increase consumer confidence in products labelled as organic, which would ensure fair competition and the proper functioning of trade in organic produce.[49]

Through Energy Star and the EU organic standard, governments have intervened with the regulatory structure of sustainability labels by setting minimum quality standards. The sustainability labels are not checked by the government but by private, third-party certifiers. The Danish organic Ø-mark and the EU quality schemes are examples of sustainability labels in which the government intervened by setting minimum quality standards that are checked by *public* bodies. As such, the government is more involved in the regulatory structure of these labels compared to Energy Star and the EU organic label.

It remains unclear, however, whether and to what extent such public or private interventions with sustainability labels are effective in increasing consumer confidence. Previous research concerning the Energy Star label[50] and the EU quality labels[51] have focused mainly on consumer willingness to pay, concluding that consumers are willing to pay a price premium for Energy Star or EU quality-labelled products. Whether a sustainability label will be effective in motivating consumers to purchase the product may not only depend on the willingness to pay, but also on consumer recognition of the label,[52] basic consumer knowledge of the production standards covered,[53] and consumer confidence in the sustainability label.[54] Previous research concerning the Danish eco-label Ø-mark has been very

[48] Energy Policy Act. Section 131 of the Act amends Section 324 (42 USC 6294) of the Energy Policy and Conservation Act.

[49] EC Regulation on Organic Agriculture, n 39, preamble para 3.

[50] Ward et al. (2011), p. 1450; Newell and Siikamäki (2013).

[51] Menapace et al. (2011), p. 193; Aprile *et al.* 2012.

[52] Thøgersen (2000), p. 285.

[53] Hoogland et al. (2007), p. 47.

[54] Thøgersen (2000), n 54; Brécard et al. (2009), p. 115.

elaborate on these three factors: the Ø-mark is recognised by almost all Danish consumers, Danish consumers seem to have a basic knowledge of the production standards,[55] and Danish consumers seem to have high confidence in their national eco-label Ø-mark compared to consumer confidence of UK, Swedish and US consumers in their national eco-label.[56] Previous research concerning the EU organic label shows low recognition of the EU organic label by EU consumers[57] and EU consumer confidence in the underlying EU organic label and the certification systems seems not very pronounced.[58] The effectiveness of these sustainability labels in increasing consumer confidence thus seems to differ. Furthermore, the Energy Star standard, the EU organic label, and the Danish Ø-mark label focus on environmental issues, so the question remains whether a similar approach would be effective and legally possible for other categories of sustainability labels (e.g. fair trade, animal welfare, health) or sustainability in general.

Further research is necessary whether and to what extent public or private actors should intervene in the regulatory structure of sustainability labels to effectively increase consumer confidence. Consumer confidence in sustainability labels may depend on the number of labels in one market and to what extent these labels address similar or different issues. To assess whether and to what extent public or private actors should intervene in the plurality of sustainability labels, future research should focus on whether and how much consumer confidence in sustainability labels is reduced with different levels of plurality of sustainability claims in one market. This can be examined in an experimental study where differences in consumer confidence caused by differences in the number of labels and issues addressed are studied. This can be done by systematically varying the extent to which consumers are confronted with a single label for each sustainability issue, or a range of labels addressing each same issue. In addition, a variation can be made whether a consumer is confronted with a single sustainability issue or a range of sustainability issues. It is expected that if multiple issues are addressed by multiple labels at the same time, consumer confidence would be lowest. Furthermore, it is expected that if multiple issues are each addressed with a single label this would lead to the highest confidence, as it gives most information with least complexity.

In addition, consumer confidence in sustainability labels may also depend on the actors in which consumers have confidence. Previous research has shown that consumers have higher confidence in sustainability labels set by NGOs[59] and that consumers do not identify or prioritise third-party certification as more independent than other types of certification.[60] Sønderskov and Daugbjerg conclude, however, that visible and substantial state involvement in standard-setting and certification

[55] Daugbjerg et al. (2014), p. 559.

[56] Sønderskov and Daugbjerg (2011), n 24.

[57] Janssen and Hamm (2014), p. 437.

[58] Janssen and Hamm (2012), p. 335.

[59] Padel and Foster (2005), n 25; Eden et al. (2008), n 24.

[60] Eden et al. (2008), n 24.

increases consumer confidence in organic labels.[61] From both studies it can be derived that consumers have low confidence in organic standards set and confirmed by firms. Both studies contradict each other whether NGOs or governments should set and confirm standards. However, they focus only on organic standards, while results may differ for other categories of sustainability. To assess whether and to what extent actors involved in standard-setting affect consumer confidence, future research should focus on whether firms, NGOs and/or governments should set and/or certify standards to increase consumer confidence and whether this may differ per dimension of sustainability (e.g. fair trade, organic, animal welfare). This could be examined with a survey where consumers are asked to compare sustainability labels on trustworthiness and willingness to pay. In this survey different sustainability labels (e.g. fair trade, organic, animal welfare) with different regulatory structures should be tested with different roles for different actors (i.e. set and/or confirmed by NGOs, governments or firms). The outcome of the experiment and the survey could be used to determine the most effective intervention strategy in sustainability labels.

This study provided an overview of the current regulatory structure underlying sustainability labels in The Netherlands. The Netherlands was used as a case study as, similar to other Western countries, it struggles with the proliferation of sustainability labels.[62] The data collection consists of publically available information as this information is also available to consumers. A limitation was that NGO-firm partnership was assumed when NGOs had public consultations with firms before standard-setting. Such a broad interpretation of NGO-firm partnership was used to not exclude NGO-firm partnership of which publically available information was ambiguous.

Private or public intervention may be necessary to effectively reduce the impact of 'bad' labels on the trustworthiness of 'good' labels. Governments may especially have a role to play as they can induce collaborations between private standard-setters and set minimum requirements that a product or production process claiming to be sustainable should meet. As governments have repeatedly stated that they wish to promote sustainable consumption,[63] the time has come for action rather than words. The options described above are, however, not the only options for governments to simulate sustainable consumption since the interpretation and application of laws may also affect sustainable consumption either positively or negatively. The next three chapters illustrate that the use of normative assumptions based on a noncritical application of neoclassical law and economics has an inhibiting effect on sustainable consumption by discouraging "good" sustainability labels (e.g. those that promote best practices in sustainable development) while encouraging "bad" sustainability

[61] Sønderskov and Daugbjerg (2011), n 24.

[62] See e.g. Koos (2011), n 1.

[63] UN General Assembly, Rio Declaration on Environment and Development, 12 August 1992, A/CONF.151/26; World Summit on Sustainable Development (2002). Plan of Implementation. Johannesburg: United Nations; UN General Assembly, *Transforming our world : the 2030 Agenda for Sustainable Development*, 21 October 2015, A/RES/70/1.

labels (e.g. those that do not promote such best practices). Insights from social sciences are used to improve the interpretation and application of these laws to better ensure sustainable consumption.

References

Abbott K, Snidal D (2009) Strengthening international regulation through transnational new governance: overcoming the orchestration deficit. Vanderbilt J Transnatl Law 501

Albersmeier F, Schulze H, Spiller A (2010) System dynamics in food quality certifications: development of an audit integrity system. Int J Food Syst Dyn 1:69

Auriol E, Schilizzi S (2003) Quality signaling through certification theory and an application to agricultural seed markets. IDEI Working Papers

Boström M (2012) A missing pillar? Challenges in theorizing and practicing social sustainability: introduction to the special issue. Sustain Sci Pract Policy 8:3

Brécard D et al (2009) Determinants of demand for green products: an application to eco-label demand for fish in Europe. Ecol Econ 69:115

Connelly S (2007) Mapping sustainable development as a contested concept. Local Environ 3:259

Daugbjerg C et al (2014) Improving eco-labelling as an environmental policy instrument : knowledge, trust and organic consumption. J Environ Policy Plan 16:559

De Boer J et al (2004) The conceptual framework. In: Eritja MC (ed) Sustainability labelling and certification. Marcial Pons

Druzin BH (2014) Anarcy, order, and trade: a structuralist account of why a global commercial legal order is emerging. Vanderbilt J Transnatl Law 47:1049

Eden S, Bear C, Walker G (2008) Understanding and (dis)trusting food assurance schemes: consumer confidence and the 'knowledge fix'. J Rural Stud 24:1

Fischer C et al (2005) Forest certification: toward common standards? Resources for the future. Discussion Paper dp-05-10. Washington, DC

Gandenberger C, Garrelts H, Wehlau D (2011) Assessing the effects of certification networks on sustainable production and consumption: the cases of FLO and FSC. J Consum Policy 34:107

Golan E et al (2001) Economics of food labelling. J Consum Policy 24:117

Grunert K, Hieke S, Wills J (2014) Sustainability labels on food products: consumer motivation, understanding and use. Food Policy 44:177

Harbaugh R, Maxwell J, Roussillon B (2011) Label confusion: the Groucho effect of uncertain standards. Manag Sci 57:1512

Hatanaka M, Bain C, Busch L (2005) Third-party certification in the global agrifood system. Food Policy 30:354

Hoogland C, de Boer J, Boersema J (2007) Food and sustainability: do consumers recognize, understand and value on-package information on production standards? Appetite 49:47

Horne R (2009) Limits to labels: the role of eco-labels in the assessment of product sustainability and routes to sustainable consumption. Int J Consum Stud 33:175

Jahn G, Schramm M, Spiller A (2005) The reliability of certification: quality labels as a consumer policy tool. J Consum Policy 28:53

Janssen M, Hamm U (2012) The mandatory EU logo for organic food: consumer perceptions. Br Food J 114:335

Janssen M, Hamm U (2014) Governmental and private certification labels for organic food: consumer attitudes and preferences in Germany. Food Policy 437

Johnston P et al (2007) Reclaiming the definition of sustainability. Environ Sci Pollut Res 14:60

Koos S (2011) Varieties of environmental labelling, market structures, and sustainable consumption across Europe: a comparative analysis of organizational and market supply determinants of environmental-labelled goods. J Consum Policy 34:127–151

Lawn P (2010) Toward sustainable development: an ecological economics approach. CRC Press, Florida

Marx A (2011) Global governance and the certification revolution types, trends and challenges. In: Levi-Faur D (ed) Handbook of the politics of regulation. Edward Elgar, Cheltenham

Menapace L et al (2011) Consumers preferences for geographical origin labels: evidence from the Canadian Olive Oil Market. Eur Rev Agric Econ 38:193

Meuwissen M et al (2003) Technical and economic considerations about traceability and certification in livestock production chains. In: Velthuis A et al (eds) New approaches to food safety economics. Kluwer Academic Publishers

Moussa M, Touzani S (2008) The perceived credibility of quality labels: a scale validation with refinement. Int J Consum Stud 32:526

Newell R, Siikamäki J (2013) Nudging energy efficiency behavior. Resources for the Future RFF Discussion Paper 13

Norgaard RB (1988) Sustainable development: a co-evolutionary view. Futures (Dec) 606

Padel S, Foster C (2005) Exploring the gap between attitudes and behavior: understanding why consumers buy or do not buy organic food. Br Food J 107:606

Raynolds LT, Murray D, Heller A (2007) Regulating sustainability in the coffee sector: a comparative analysis of third-party environmental and social certification initiatives. Agric Hum Values 24:147

Redclift M (1992) The meaning of sustainable development. Geoforum 25:395

Sønderskov KM, Daugbjerg C (2011) The state and consumer confidence in eco-labelling: organic labelling in Denmark, Sweden, the United Kingdom and the United States. Agric Hum Values 28:507

Teisl M et al (2002) Consumer reactions to environmental labels for forest products: a preliminary look. For Prod J 52:44

Thøgersen J (2000) Psychological determinants of paying attention to eco-labels in purchase decisions: model development and multinational validation. J Consum Policy 23:285

Van Dam Y, De Jonge J (2015) The positive side of negative labelling. J Consum Policy 38(1): 19–38

van Dam YK, van Trijp HCM (2011) Cognitive and motivational structure of sustainability. J Econ Psychol 32:726

Van der Zee E (2018) Disciplining private standards under the SPS and TBT Agreement. A plea for market-state procedural guidelines. J World Trade 52

Ward D et al (2011) Factors influencing willingness-to-pay for the ENERGY STAR Label. Energy Policy 39:1450

Wilkinson J (2007) Fair trade: dynamic and dilemmas of a market oriented global social movement. J Consum Policy 30:219

Chapter 3
Sustainability Labels in the Shadow of EU Consumer Law

3.1 Introduction

Green pictograms are pictorial symbols, using visual information commonly associated with organic production, e.g. the colour green and images of leaves.[1] By using this visual information, green pictograms may give consumers the impression that the product it is attached to is produced organically. Not all green pictograms guarantee such production, however. Organic production and independent checks to guarantee organic production may warrant a price premium to consumers.[2] To gain the price premium without the higher production and certification costs, food businesses may be incentivized to free-ride on those pictograms that guarantee organic production. Such practice may give consumers the impression that the food product is produced in an organic manner; which may be considered misleading in the European Union (EU) under the Food Information Regulation (FIR).[3]

In the EU, the FIR aims at forming a framework for EU and national measures to ensure that consumers are not misled when buying food products. To assess whether commercial practices, such as green pictograms, are misleading *in concreto*, meaning in the case at hand has to be assessed by national courts against the normative benchmark of the "average consumer", who is "reasonably well-informed and reasonably observant and circumspect".[4] The "average consumer" benchmark is a

---

This book chapter is based on Eva van der Zee and Arnout Fischer, 'Green Pictograms on EU Foods: A Legal Study Informed by Behavioural Science' (2018) 7(1) Journal of European Consumer and Market Law, pp. 15–22.

---

[1] Hahnel et al. (2015), Chap. 1, n 1.

[2] Cafaggi and Iamiceli (2013), p. 136.

[3] EU Food Information Regulation, Chap. 1, n 57.

[4] See e.g., Case 315/92 *Clinique*, ECLI:EU:C:1994:34 and Case 210/96 *Gut Springheide*, ECLI:EU:C:1998:369.

© The Author(s), under exclusive license to Springer Nature Switzerland AG 2022

E. van der Zee, *Sustainability Labels in the Shadow of the Law*, Studies in European Economic Law and Regulation 22, https://doi.org/10.1007/978-3-030-95802-2_3

normative test developed by the European Court of Justice (ECJ). The ECJ has clarified what can be expected from this normative "average consumer" when confronted with visual information on consumer products, but only when the visual information was corrected by textual information.[5] It is not obvious from ECJ case-law what can be expected from the normative "average consumer" when confronted with visual information that is not corrected by textual information, such as pictograms. Behavioural research may provide insights as to how the "real consumer", confined by the methodology, processes visual information. As such, behavioural research may be helpful to inform judicial decision making as to what can be expected from the normative "average consumer". Clarity is required as to whether and how behavioural insights could guide national courts in the doctrinal assessment of misleading visual information by national courts.

The objective of this study is to examine whether and how insights from behavioural sciences could inform the doctrinal assessment of the normative EU "average consumer" benchmark under the FIR when confronted with purely visual information as used by green pictograms. To achieve the objective two pictograms that exist on food products in the Dutch market that use visual properties that consumers may associate with organic production were selected, with only one pictogram guaranteeing organic production. It was, first, examined doctrinally whether and to what extent the FIR could be interpreted and applied to the pictogram that does not guarantee organic production.[6] Second, as the analysis under the FIR is based on the normative benchmark of the "average consumer", the expectations of the ECJ of the normative "average consumer" were examined. Third, it was analysed whether insights from behavioural science could inform judicial decision making under the FIR. Subsequently, two experiments were conducted to assess the potential benefits of including behavioural research under the normative "average consumer" test. Finally, the results from the legal and experimental study were compared to assess the usefulness of behavioural research in the legal analysis of the normative benchmark of the "average consumer" under the FIR.

3.2 Green Pictograms and EU Law

The EU Regulation on Organic Agriculture defines the objectives, principles and rules applicable to organic production to contribute, amongst others, to consumer confidence.[7] The EU Regulation on Organic Agriculture seeks to solve the regulatory dilemma of organic production by combining best environmental practices, a high level of biodiversity, the preservation of natural resources, the application of high animal welfare standards and a production method using natural substances and processes.[8] When producers comply with this Regulation, they may use textual

[5]Case 470/93 *Mars*, ECLI:EU:C:1995:224; Case 195/14 *Teekanne*, ECLI:EU:C:2015:361.

[6]Cf Hutchinson (2015), Chap. 1, n 54.

[7]Preamble 5, EC Regulation on Organic Agriculture, Chap. 2, n 39.

[8]Ibid Preamble 1.

Fig. 3.1 Shop pictogram (By using this shop label, it is in no way implied that the Dutch retail chain using this quality stamp is doing anything illegal)

Fig. 3.2 EU pictogram

information such as "organic", "eco", and "bio",[9] and/or the EU pictogram (Fig. 3.2) on their food packaging.[10] The use of textual information such as "organic", "eco", and "bio" is not allowed when producers do not comply with the EU Regulation on Organic Agriculture.[11] The use of green pictograms using similar visual properties as the EU pictogram is, however, not covered by the EU Regulation on Organic Agriculture.[12]

Similar to the terms "organic", "eco", and "bio", green pictograms may give consumers the impression that a food product is produced using organic production methods. An example of a green pictogram that may give consumers such an impression is the shop pictogram shown in Fig. 3.1. Similar to the EU pictogram (Fig. 3.2), the shop pictogram uses visual properties that consumers may associate with organic production: the colour green, an image of a leaf, and the form of a quality stamp. In the consumer's mind, both pictograms may create the impression that a product carrying any of these stamps is produced organically, while this impression is only factually correct for the EU pictogram. The EU pictogram is certified by an independent organization to comply with the EU Regulation on Organic Agriculture,[13] while the shop pictogram is a mere image without any underlying standards. As such, the shop pictogram may be considered to give consumers the misleading impression that the product is produced organically.

To assess whether the shop pictogram could be considered misleading, the FIR would be applicable. In the EU, the FIR aims at forming a legal framework for EU and national measures to ensure that consumers are not misled when buying food products. The FIR complements the Unfair Commercial Practices Directive (UCPD)[14] which covers certain aspects of the provision of information to consumers

---

[9]Ibid art 23.

[10]Ibid art 25.

[11]Ibid art 23.

[12]Preamble 5, EC Regulation on Organic Agriculture, Chap. 2, n 39.

[13]Ibid.

[14]Council Directive (EC) 2005/29 concerning unfair business-to-consumer commercial practices in the internal market [2005] OJ L 149 (Unfair Commercial Practices Directive).

specifically to prevent misleading actions and omissions of information.[15] Both the UCPD and the FIR take the normative "average consumer" as clarified by the ECJ as a normative benchmark to examine whether consumers are misled.[16]

The FIR states that food law shall provide a basis for consumers to make informed choices about the food they consume and to prevent any practice that may mislead the consumer (recital 4 FIR). The FIR requires that food information must be fair, meaning that it may not be misleading (Article 7 FIR). More specifically for our case at issue, food information may not be unfair "as to the characteristics of the food and, in particular, as to its nature, identity, properties, composition, quantity, durability, country of origin or place of provenance, method of manufacture or production" (Article 7(1)(a) FIR) or "by attributing to the food effects or properties which it does not possess" (Article 7(1)(b) FIR). Furthermore, the FIR gives some weight to how consumers process information by emphasising that food information shall be accurate, clear and easy to understand for the consumer (Article 7(2) FIR), i.e. the normative "average consumer" as clarified by the ECJ.

When a pictogram gives the normative "average consumer", as clarified by the ECJ, the misleading impression that the food product it is attached to is produced organically, the pictogram could be considered misleading under the FIR. Unlike other rules of EU consumer protection legislation,[17] Article 7 FIR does not explicitly require a causal link between the misleadingness of the food information and the transactional decision of the normative "average consumer". Some scholars argue that general legal requirements and good scientific practice of consumer decision theory require that a study examining whether product labels could be misleading needs to include an examination of whether the misleadingness affects the transactional decision made by the normative "average consumer".[18]

3.3 The Normative Benchmark of the "Average Consumer" Based on ECJ Case Law

To assess whether visual information as used by green pictograms is misleading under the FIR should be based on what can be expected from the normative "average consumer" benchmark when confronted with such visual information. This normative benchmark as developed by the ECJ gives guidance to national courts and the

[15] Preamble 5 FIR.

[16] Unfair Commercial Practices Directive, n 14, preamble 18, arts 6–8. See also Case 195/14 *Teekanne*, ECLI:EU:C:2015:361, n 5, at 36, where the ECJ applied the "average consumer" benchmark to assess the misleadingness of food information under the under the Directive preceding the FIR (Council Directive (EC) 2000/13 on the approximation of the laws of the Member States relating to the labelling, presentation and advertising of foodstuffs [2000] OJ L 109).

[17] Unfair Commercial Practices Directive, n 14, art 6(1).

[18] Purnhagen and Van Herpen (2017), p. 229.

national legislator. It is not for the ECJ to rule on the question of whether the labelling of certain products is likely to mislead the consumer; this is a task for the national courts.[19] The normative benchmark of the "average consumer" was first introduced by the ECJ in 1998 in Gut Springheide.[20] Gut Springheide concerned a company that marketed eggs ready-packed under the description '6-Korn – 10 frische Eier' (six-grain – 10 fresh eggs). The company asserted that the six varieties of cereals in question account for 60% of the feed mix used to feed the hens. A slip of paper enclosed in each pack of eggs praised the beneficial effect of this feed on the quality of the eggs. According to the German Courts, the information provided on and in pack would be prohibited under the German Foodstuffs and Consumer Goods Law. Gut Springheide asserted that the German appeal court did not produce expert opinion to prove that they misled the consumer. The case moved its way up to the Bundesverwaltungsgericht (the Federal Administrative Court) who asked the ECJ for a preliminary ruling. The Bundesverwaltungsgericht asked the ECJ whether misleadingness should be determined based on the actual expectations of consumers to whom the information is addressed or based on an objectified concept of a consumer open only to legal interpretation.

The ECJ referred to its judgments in the interpretation of the application of the free movement of goods[21] and held that to determine whether a description, trade mark or promotional description or statement could be considered misleading should be assessed against the "average consumer" who is "reasonably well-informed and reasonably observant and circumspect".[22] The ECJ expected that national courts should, in general, be able to assess a misleading effect of a description or statement designed to promote sales, based on this normative "average consumer" benchmark, without ordering an expert's report or commissioning a consumer research poll.[23] In certain circumstances, however, e.g. when the necessary information is not at the disposal of the court or where the solution was not clear from the information before it, the ECJ held that a national court "may have recourse to a consumer research poll or an expert's report as guidance for its judgment".[24]

For national courts to apply the normative benchmark of the "average consumer" to green pictograms, then depends on the question how well-informed and observant the normative "average consumer" can reasonably be considered to be. Acknowledging that many cases influenced the interpretation of the normative

[19] See, in particular, Case 366/98 *Geffroy*, EU:C:2000:430 at 18–20, and Case 446/07 *Severi*, EU:C:2009:530 at 60.

[20] Case Case 210/96 *Gut Springheide*, ECLI:EU:C:1998:369.

[21] In particular Case 362/88 *GB-INNO-BM*, ECLI:EU:C:1990:102; Case 238/89 *Pall*, ECLI:EU:C:1990:473; Case 126/91 *Yves Rocher*, ECLI:EU:C:1993:191; Case 315/92 *Clinique*, ECLI:EU:C:1994:34; Case 456/93 *Langguth*, ECLI:EU:C:1995:78 and Case 470/93 *Mars*, ECLI:EU:C:1995:224, n. 131.

[22] Case Case 210/96 *Gut Springheide*, ECLI:EU:C:1998:369, at 30–32.

[23] Ibid at 31.

[24] Ibid at 37.

benchmark of the EU "average consumer",[25] the three leading cases that shaped the legal expectations of the "average consumer" in EU law towards food labelling will be focused on in this chapter.

In earlier case-law, most notably the *Mars*-case, the ECJ explained what a reasonably circumspect consumer would be considered to know when confronted with potentially misleading visual information front of pack accompanied by correcting textual information positioned next to the visual information. The *Mars*-case concerned a publicity campaign by Mars inc. in which the wrapping of ice-cream bars depicted a coloured band bearing the textual information "+ 10%". The visual information, the coloured band, was much larger than indicated by the textual information. The plaintiffs to the case at the German Court contended that a significant number of consumers "will be induced into believing [...] that the increase is larger than represented" and that, therefore, the marketing should be prohibited.[26] In providing an answer to the *Landgericht Köln* as to whether such a prohibition is compatible with the internal market, the ECJ referred to the normative benchmark of the "average consumer". The ECJ held that "[r]easonably circumspect consumers may be deemed to know that there is not necessarily a link between the size of publicity markings relating to an increase in a product's quantity and the size of that increase".[27] Without explicitly stating this, it seems that the ECJ considered the textual information to be sufficiently determinate to consider that a "reasonably circumspect consumer" would not misinterpret the visual information on the food packaging.

Whether under the normative benchmark of the "average consumer", the "average consumer" can be considered "reasonably well-informed and reasonably observant and circumspect" with relation to textual information on food packaging was further elaborated by the ECJ in Adolf *Darbo*.[28] The case concerned a jam manufacturer under the name "Darbo Naturrein" (naturally pure) and under the description "Garten Erdbeer" (garden strawberry). A German consumer organization argued that because the jam contained the additive pectin and traces or residues of lead, cadmium, and pesticides, the term "Naturrein" (naturally pure) and "Garten Erdbeer" (garden strawberry) was likely to "create in the consumer's mind the impression that Darbo jam is a pure and natural product, free of any impurity or extraneous substance".[29] The *Oberlandesgericht Köln* sought the advice of the ECJ, whether the name "Naturrein" was allowed by Directive 79/112/EEC (on the labelling, presentation and advertising of foodstuffs).[30] The ECJ held that the normative

[25] Cf. Duivenvoorde (2015), pp. 29–52; Mak (2011), pp. 27–29.

[26] Case 470/93 *Mars*, ECLI:EU:C:1995:224, n. 131, at 22.

[27] Ibid at 24.

[28] Case 465/98 *Darbo*, ECLI:EU:C:2000:184.

[29] Ibid at 26.

[30] Council Directive (EEC) 79/112 on the approximation of the laws of the Member States relating to the labelling, presentation and advertising of foodstuffs for sale to the ultimate consumer, OJ L 33.

"average consumer" would not be misled by the term "Naturrein" on the label, because the additive pectin was listed in the list of ingredients.[31] Furthermore, the ECJ held that it is common knowledge that garden fruit is inevitably exposed to pollutants.[32] This judgment by the ECJ could be understood as that the normative "average consumer" is expected to read the list of ingredients before purchasing a product and to have a general knowledge that environmental pollution may affect food production.

More recently, the ECJ has clarified what can be expected from the normative "average consumer" when potentially misleading visual and textual information front of pack is corrected by textual information that is positioned back of pack. *Teekanne* is a tea producer who marketed fruit tea under the name "Felix Himbeer-Vanille Abenteuer" (Felix raspberry and vanilla adventure). The packaging contained depictions of raspberries and vanilla flowers, the indication "Früchtetee mit natürlichen aromen" [sic] (fruit tea with natural flavourings) and "Früchtetee mit natürlichen aromen – Himbeer-Vanille-Geschmack" (fruit tea with natural flavourings – raspberry-vanilla taste), and a seal with the indication "nur natürliche Zutaten" (only natural ingredients). The ingredient list on the side of the packaging indicated, however, that the fruit tea did, in fact, not contain any vanilla or raspberry constituents or flavourings. The *Bundesgerichtshof* referred to the ECJ whether it is permissible under the Directive preceding the FIR[33] for the labelling, presentation and advertising of foodstuffs to give the impression, through their appearance, description or pictorial representation, that a particular ingredient is present, even though that ingredient is not in fact present and this is apparent solely from the list of ingredients. The ECJ held that textual information on the back of a food package, such as an ingredient list, does not sufficiently counter a "consumer's erroneous or misleading impression" caused by visual and textual information on the front of that food package.[34] Therefore, the ECJ concluded that the food label "taken as a whole" should be taken into account by national courts to assess whether the normative "average consumer" is misled as to the characteristics of the food.[35]

Taken together the views by the ECJ on how national courts could assess the misleading effect of information on food packaging mainly gives guidance to national courts as to how to examine the misleadingness of either solely textual information or visual information accompanied by textual information. Most assumptions of the normative "average consumer" test are based on assumptions of the normative "average consumer" when confronted solely with textual information.[36] In the two cases presented above, where the normative "average consumer"

[31] Case 465/98 *Darbo*, ECLI:EU:C:2000:184., n 28, at 22.

[32] Ibid at 27.

[33] Council Directive (EC) 2000/13, n 16.

[34] Case 195/14 *Teekanne*, ECLI:EU:C:2015:361, n 5, at 40.

[35] Ibid at 41–2.

[36] This is also the case for commercial information that does not appear on food packaging. See Case 362/88 *GB-INNO-BM*, ECLI:EU:C:1990:102, n 21; Case 126/91 *Yves Rocher*, ECLI:EU:C:1993:

test was applied to visual information, this visual information was always accompanied by correcting texts.[37] In the *Mars* case, the visual information was accompanied by correcting texts front of pack. In the *Teekanne* case, the visual and textual information front of pack were corrected by textual information back of pack. None of these cases involved visual information in the complete absence of correcting texts.

When national courts want to examine whether pictograms that are not accompanied by textual information, such as the shop pictogram, is misleading, the normative benchmark of the "average consumer" may not give sufficient guidance. To get a better understanding of what can be expected from the normative "average consumer" when confronted with green pictograms, insights from behavioural research may be helpful to inform judicial decision making.

3.4 The Normative Benchmark of the "Average Consumer" and Behavioural Research

It remains unclear whether behavioural research could inform judicial decision making in EU consumer law as Recital 18 UCPD emphasizes that the normative benchmark of the "average consumer" is not a statistical test.[38] It remains unclear how Recital 18 UCPD should be understood. A very strict interpretation of Recital 18 may be that because the UCPD aims at ensuring the proper functioning of the internal market and to protect consumers' economic interests (Article 1 UCDP), a normative decision should be taken that is not based on any statistical data. This interpretation leaves open whether Recital 18 UCPD has an impact on the normative benchmark of the "average consumer" assessed in the FIR. The FIR is a *lex specialis* to Article 8 General Food Law (GFL)[39] and therefore not necessarily subjected to the UCPD.[40] While the UCPD is more market-oriented, the GFL is more consumer-oriented, as it aims to provide a high level of protection of human health and consumers' interest about food (Article 1 GFL). Following this argument, to ensure

191, n 21; Case 210/96 *Gut Springheide*, ECLI:EU:C:1998:369; Case 220/98 *Lifting*, ECLI:EU:C:2000:8; Case 239/02 *Douwe Egberts v Westrom Pharma*, ECLI:EU:C:2004:445; Case 26/13 *Kásler*, ECLI:EU:C:2014:282.

[37] Case 470/93 *Mars*, ECLI:EU:C:1995:224, n. 131; Case 195/14 *Teekanne*, ECLI:EU:C:2015:361, n 5.

[38] Howells et al. (2006), p. 116; Trzaskowski (2016), p. 33.

[39] Regulation (EC) 178/2002 laying down the general principles and requirements of food law [2002] OJ L 31 (General Food Law)

[40] Unfair Commercial Practices Directive, n 14, recital 5 and art 3(4).

such high consumer protection, statistical research may provide useful insights within the framework of the FIR regardless of Recital 18.[41]

Another interpretation of Recital 18 may be that national courts do not have to use empirical research to come to a normative conclusion. In our view, such interpretation would be the most accurate. Although the ECJ held that national courts should be able to assess on their own whether the normative "average consumer" can be considered misled,[42] the ECJ explicitly stated that a national court may nonetheless decide to order an expert's opinion or commission a consumer research poll.[43] To extrapolate a poll to more general insights that are useful in the assessment of whether the normative "average consumer" could be considered misled, statistics are essential. As such, the judgment by the ECJ suggests that at least some type of statistical evidence is allowed. This is also in line with the general trend in the case-law of the ECJ to leave open the possibility for Member States to use statistical evidence or any other suitable type of evidence to justify measures infringing the free movement of goods, where the "average consumer" benchmark is part of.[44] This could be considered an opening towards including empirical studies into the actual behaviour of consumers. In the case of the current chapter, empirical research may provide insights into a legal analysis of whether the shop pictogram gives the normative "average consumer" the impression of sustainability.

3.5 Experiments

To examine whether and how insights from behavioural science could inform the doctrinal assessment of the normative EU "average consumer" benchmark as clarified by the ECJ under the FIR to purely visual information as used by pictograms, two experiments were conducted. These experiments serve as a showcase to examine how insights from behavioural research could inform judicial decision making of pictograms under the FIR. The following experiments could be used by national courts to examine the misleadingness of shop pictograms that closely resemble official pictograms. We investigated consumer interpretation of the shop pictogram and the EU pictogram (study 1) applied to product packaging (study 2). To analyse this, we, first, investigated whether a shop pictogram using similar visual properties as the official EU pictogram, the colour green, an image of a leaf, and the form of a quality stamp, gives consumers the impression of a higher level of organic

[41] It should be noted, however, that such interpretation fails to understand that statistical research could add to the information base of the judge to come to a more well-founded normative decision. See e.g. Trzaskowski (2016), n 38.

[42] Case Case 210/96 *Gut Springheide*, ECLI:EU:C:1998:369, at 33–35.

[43] Ibid at 31.

[44] Case 333/14 *Scotch Whiskey*, ECLI:EU:C:2015:845, at 64–5 in conjunction with Case 148/15 *Deutsche Parkinson Vereinigung*, ECLI:EUC:2016:776, at 35–6.

production compared to no logo. Second, we examined whether this impression is similar to the impression given by the official logo. Third, we examined whether simultaneous exposure to both pictograms would correct the consumer's misleading impression of the shop pictogram. To get as close to reality as possible these issues were not only investigated with the pictograms on their own but also applied to a product.

In the first study, it was specifically assessed whether participants are likely to infer organic properties from the pictograms analysed or whether they were likely to (also) infer other properties, i.e., fair trade, animal welfare, free of (artificial) colourings and flavourings, or health. In the second study, it was studied whether participants also infer these properties when the pictograms were attached to a food product and whether this would lead participants to be willing to make a larger monetary sacrifice to obtain such a product.

3.5.1 *Study 1*

3.5.1.1 Design and Procedure

A convenience sample of 70 Dutch participants consisting of students in an undergraduate course and people living close to the home address of the first author volunteered for the experiment. Age ranged between 18 and 35 ($M = 22$, SD 4.15). Of the participants, 43 were female, 26 were male and 1 participant did not report gender. The experiment had a two-group design (shop pictogram and EU pictogram—see Figs. 3.1 and 3.2 respectively).

Participants were invited to follow a link to fill in the survey in their own time, on their own computer or smartphone. After clicking the link, participants were randomly assigned to one of the conditions. Depending on the condition, participants were shown a picture of either the EU pictogram or the shop pictogram. Participants were asked to rate to what extent they agreed that a product with that specific label was (1) Fair Trade (2) Animal Friendly (3) Organic (4) Free from artificial flavours and colorants (5) Healthy, on a visual analogue scale coded from 0 to 100 (0 = completely disagree; 100 = completely agree). All participants were then shown both pictograms together and scored for each pictogram to what extent participants perceived the product to be organically produced on the same visual analogue scale as before. Completing the experiment took about 3 min.

3.5.1.2 Results

An independent t-test showed no differences between the pictograms for participants rating of organic $t(68) = 0.71$; $p = .48$ (Table 3.1). Participants scored the pictograms highest on indicating organic properties, although in the case of the EU pictogram the difference with fair trade ($p = .15$) and healthiness ($p = .07$) was

Table 3.1 Means and standard deviations for the statements

	EU pictogram			Shop pictogram			
	n	Mean	SD	*n*	Mean	SD	Cohen's d
On its own							
Organic	35	69.57[a]	26.27	35	73.94[a]	25.41	0.17
Fair trade	35	60.03[ab]	29.86	35	44.37[c]	33.67	0.49
Animal friendly	35	57.17[b]	27.73	35	42.74[c]	30.76	0.49
Free of artificial additives	35	50.86[b]	26.11	35	60.60[b]	29.92	0.34
Healthy	35	59.71[ab]	23.72	35	61.20[b]	30.41	0.05
In comparison							
Organic	70	67.74	26.06		60.91	27.53	0.23

Notes: Cohen's d measure for effect size. Around 0.20 small effects, about 0.50 medium effects (Cohen 1988). Means in columns that share a superscript character are not significantly different at $p = .05$ (pairwise comparisons repeated measures least significant difference (LSD))

not significant. Participants considered the EU pictogram to be indicative of a higher level of fair trade $t(68) = -2.06$; $p = .04$, and animal friendliness $t(68) = -2.06$; $p = .04$; neither of which are regulated by the EU pictogram. No differences in rating for free of artificial additives $t(68) = 1.45$; $p = .15$ or healthiness $t(68) = -23$; $p = .82$ were found. When participants were shown both pictograms, a paired sample t-test showed no differences for perceived level of organic $t(69) = -1.27$; $p = .21$. Hence, this study suggests that participants perceived the pictograms as an indication of organic production. In addition, the study shows no difference in the extent to which participants perceive the labels to indicate organic production.

3.5.2 *Study 2*

3.5.2.1 Design and Procedure

Participants were 122 students from Wageningen University between 17 and 26 years ($M = 21$, SD $= 2.0$), of whom 80 (66%) were female. Students were approached midterm during the fall semester of 2016 in an education building and were asked to participate in a series of studies. After completing the first, unrelated study they participated in the current study which took about 5 min. As a reward for the combined study, which lasted about 20 min, they received a snack representing a monetary value of about €1.00.

Three different images of Dutch syrup waffles (*stroopwafels*) were prepared. Images either showed no pictogram (Fig. 3.3a), the shop pictogram (Fig. 3.3b), or the EU pictogram (Fig. 3.3c). Participants were shown only two of the packages which were presented next to each other. This allowed direct comparison. Participants were asked to make one of three comparisons: (1) No pictogram—vs—shop pictogram, (2) No pictogram—vs—EU pictogram, and (3) shop pictogram—vs—EU pictogram.

Fig. 3.3 The three syrup waffles packages presented to participants. Each participant saw only 2 out of these 3 packages. (**a**) No pictogram. (**b**) Shop pictogram. (**c**) EU pictogram

Table 3.2 Means (SE) of aggregated scores for the three products on the perceived product property

	No label	Shop pictogram	EU pictogram
Organic	20.83 (1.97)^{aII,III}	58.64 (3.21)^{bI}	64.45 (2.93)^{bI}
Fair trade	23.46 (2.28)^{aII}	43.83 (3.28)^{bII}	49.74 (3.30)^{bII}
Animal friendly	29.75 (2.56)^{aI}	42.81 (2.94)^{bII}	45.53 (3.35)^{bII}
Free of artificial additives	24.06 (2.11)^{aII}	46.55 (2.84)^{bII}	50.48 (2.60)^{bII}
Healthy	15.46 (1.88)^{aIII}	18.54 (2.05)^{aIII}	17.58 (2.13)^{aIII}

Values within a row that share a superscript character or within a column that share a roman numeral are not significantly different at $p = .05$ (pairwise comparisons LSD)

The same questions about the perceived level of organic, fair trade, animal friendliness, free of artificial additives and healthiness as in study 1 were asked. In addition, willingness to pay (WTP) for each package was assessed. After giving them the information that similar packages of syrup waffles are sold in retail for €1.50, participants were asked to give a price indication of what they would be willing to pay for these packages in Euros up to two decimals.

3.5.2.2 Results

Overall scores on perceptions of organic, fair trade, animal friendliness, freedom of artificial additives and healthiness for the packages are given in Table 3.2. Different from study 1, no differences between the shop pictogram-labelled product and EU pictogram-labelled product for fair trade and animal friendliness were found. Similarly, to study 1, it was found that participants mainly inferred organic properties from the labels. Both the shop pictogram-labelled product and the EU pictogram-labelled product were rated higher on organic, and the other sustainability properties fair trade, animal friendliness, and free of acritical additives than the package without a pictogram. This confirmed the suggestion that the pictograms convinced

Table 3.3 Perceived level of organic (Mean, (SD)) of a product with a label presented next to another product

Condition	N	No label	Shop pictogram	EU pictogram	Paired t, p, Cohen's d
No label—shop label	40	20.25 (17.04)	61.88 (28.55)		t = 8.82, p < .01, d = 1.46
No label—EU pictogram	41	21.39 (18.55)		68.66 (25.59)	t = −8.76, p < .01, d = 1.38
Shop pictogram—EU pictogram	41		55.49 (29.20)	60.24 (27.10)	t = −0.81, p = .42, d = 0.13

Notes: Cohen's d around 0.20 small effects, about 0.50 medium effects, around 0.80 and above large to very large (Cohen 1988)

participants that the associated product was more organic than conventional products. No differences were found between the perception of healthiness across the labelled and unlabelled products, with all cases of this sweet pastry scoring low on the healthiness scale.[45]

Overall scores on perceptions of organic for the packages in comparison are given in Table 3.3. Paired sample t-tests on both the products rated by the same participant showed no significant difference between the shop-labelled product and the EU-labelled product $t(40) = -0.81, p = .42$. Differences between the shop-labelled product and the non-labelled product $t(39) = 8.82$, $p<.01$, and the EU-labelled product and the non-labelled product $t(40) = -8.76$, $p < .01$ were found. This provides additional support for the findings from study 1 that either pictogram suggested that the product was more organic than the product without a pictogram and that no differences between the two pictograms could be identified.

Next, it was explored whether organic perception of the labels influenced economic decision making of participants. It was found that participants are willing to pay approximately 28 Eurocents (25%) more for the shop pictogram-labelled product compared to the product without pictogram $t(39) = 3.48$, $p < .01$, about 20 Eurocents (14%) for the EU-labelled compared to the product without pictogram $t(40) = 3.66$, $p < .01$. The difference of 8 Eurocents (5%) between the shop pictogram and the EU pictogram did not reach significance $t(40) = -1.92$, $p = .06$ (Table 3.4). Hence, these results suggest participants were willing to sacrifice substantially more money for either pictogram-labelled products than for the product without pictogram.[46]

[45] It should be noted that as the pictograms were attached to a sweet pastry (stroopwafels) product, consumers may associate the pictograms less with health, than if they were applied to a healthier product (e.g. pasta sauce).

[46] It was also found that willingness to pay is moderately (0.37) to strongly (0.95) positively correlated between the rated products. This implies that participants willing to pay more for one of the packages is also willing to pay more for the other, which, unsurprisingly implies that willingness to pay depends on liking of syrup waffles in general.

Table 3.4 WTP in € (mean, (*SD*)) of a product with a label presented next to another product

Condition	*n*	No label	Shop pictogram	EU pictogram	Paired *t*, *p*, Cohen's *d*
No label—shop label	40	1.14 (0.41)	1.42 (0.47)		$t = 3.48, p < .01, d = 0.57$
No label—EU pictogram	41	1.42 (1.11)		1.62 (1.10)	$t = -3.66, p < .01, d = 0.57$
Shop pictogram—EU pictogram	41		1.34 (0.61)	1.42 (0.61)	$t = -1.92, p = .06, d = 0.38$

Notes: Cohen's *d* around 0.20 small effects, about 0.50 medium effects, around 0.80 and above large to very large (Cohen 1988)

3.5.3 Discussion

Study 1 showed that participants associated the shop pictogram and the EU pictogram mainly with organic production; which suggests consumers perceive either pictogram as an indication of organic production. Study 2 confirmed that both the shop and the EU pictogram gave participants the impression that the product is more organic than a product without a pictogram as both pictogram-labelled products score substantially higher on organic than the non-labelled product. Hence, it can be concluded that consumers infer the property organic from either pictogram, while only the EU pictogram guarantees organic production.

In both studies, participants rated the level of organic of the EU pictogram similar to that of the shop pictogram. This was the case when the pictograms were presented in isolation and when they were brought in direct comparison (study 1), and when they were presented on a package and shown in direct comparison (study 2). Caution is needed not to over-interpret non-significant difference as an indication of no difference. An alternative explanation can be a lack of power, i.e., that there are too few observations to make a definitive claim or other methodological shortcomings. In the current case, studies 1 and 2 showed that differences as to the perceived level or organic between labelled products is at best small,[47] while the difference between the labelled and the non-labelled product was much more substantial. This suggests that both pictograms convey a similar message about how organic the product is produced and that the presence of the EU pictogram does not sufficiently correct the consumer's misleading impression of the shop pictogram. Furthermore, the message these pictograms conveyed have behavioural and economic consequences for participants as participants reported a higher willingness to pay a price premium for the labelled products compared to the non-labelled product.

Both studies showed that besides the property organic, also other sustainability properties were positively influenced by the pictograms. In both studies fair trade, animal friendly, and free of artificial additives were rated relatively high, albeit

[47] When interpreting Cohen's d, a measure for effect sizes independent of the number of observations.

slightly lower scores than the property organic. This is consistent with consumer studies showing that people tend to see sustainability as a unified construct that combines organic, fair trade and other sustainable properties.[48] Health scores were substantially higher for the labelled products in study 1, but not different from non-labelled products in study 2. This is probably because a health inference is unlikely for the sweet pastry the pictograms were applied to in study 2.

Before discussing the use of these insights for the normative benchmark of the "average consumer" in the FIR, the limitations of the current approach should first be considered, i.e., mainly to what extent it is possible to generalize consumer studies to the "real world".[49] Firstly, in the reported studies participants were not a representative sample of the Dutch population. In study 1 half of the participants were students of Wageningen University, the other half were graduates from Utrecht University. In study 2, all participants were students of Wageningen University. Students and university graduates can be expected to have a higher level of education and to be more interested in organic production than the general population. Thus, the sample has likely more knowledge about green pictograms due to a higher motivation to choose products produced organically. For such a knowledgeable and highly educated sample more critical distinction between the shop pictogram and the EU pictogram would be expected. Nevertheless, the perception and valuation of the shop pictogram and the EU pictogram were very similar, which makes it likely that the difference in perception and valuation of the shop pictogram compared to the EU pictogram will be even smaller in the general population.

Secondly, the surveys were conducted in front of a computer screen which is different from a shop setting where consumers would normally be confronted with such pictograms. While in a supermarket consumers awareness of pictograms is limited, when asked to participate in a survey participants are probably more aware of the pictograms.[50] This increased awareness of participants would be expected to result in a more critical evaluation of the shop pictogram compared to the EU pictogram than in a supermarket. Therefore, it is likely that the distinction between the shop pictogram and the EU pictogram in a real shopping environment is even smaller than in the current study.

The reported studies were done with only Dutch participants and it may well be that consumers in other countries differently value EU directives, and therefore may differently appreciate the EU pictogram. For example, English consumers are shown to be more sceptical than other European consumers to trusting commercial or EU information.[51] The ECJ already held that "social, cultural and linguistic factors"[52] should be taken into account to determine whether the normative "average

---

[48] Van Dam and Van Trijp (2011), Chap. 2, n 17.

[49] Cf Hillman (2000), p. 717; Rachlinski (2007), p. 739.

[50] Cf Grunert and Wills (2007), p. 396.

[51] Miles et al. (2005), p. 253.

[52] See e.g., Case 315/92 *Clinique*, ECLI:EU:C:1994:34 and Case 210/96 *Gut Springheide*, ECLI:EU:C:1998:369.

consumer" could be considered misled. To take such "social, cultural and linguistic factors" into account, the use of national samples to determine whether the normative "average consumer" will be misled *in concreto*, may be appropriate to provide insights to the doctrinal assessment. Using a national sample may especially be appropriate in our case at issue, where both products and one of the pictograms are typically Dutch. Although the specific outcomes are limited to the Netherlands, the current study may suggest ways how other national courts with the broader EU framework may contract behavioural research on a case by case basis.

Concluding, the conducted experiments provide support for the argument (at least when applied to a Dutch setting) that under the FIR the normative "average consumer" may be considered misled when confronted with the shop pictogram.

3.6 Conclusion and Discussion: The "Average Consumer" Test and the Behavioural Study

Whether pictograms on food packaging could be considered misleading under the FIR, revolves around the question of what can be expected from the normative "average consumer". Drawing on Article 7 FIR, the behavioural study provides guidance to the normative benchmark of the "average consumer" by indicating that the shop pictogram likely leads the normative "average consumer" to incorrectly infer characteristics of the product and by attributing properties to the food it does not possess, which is prohibited under Article 7(1)(a) and Article 7(1)(b). Furthermore, as participants assign meaning to the shop pictogram that goes beyond the mere image of a green leaf but adopts properties of pictograms that guarantees organic production, our results indicate that the shop pictogram may be interpreted as not being accurate, clear and easy to understand by the normative "average consumer", while this is required under Article 7(2) FIR. Furthermore, participants were willing to pay more for the shop pictogram-labelled product than for a non-labelled product and valued a shop pictogram-labelled product and the EU pictogram-labelled product at a comparable price. It would be unlikely that the normative "average consumer" would have taken such a transactional decision if they were aware that the shop pictogram does not imply any level of sustainable production. These results are, therefore, an indication that there is a causal link between the misleadingness of the shop label and the transactional decision of the consumer,[53] which further strengthens the conclusion that the shop pictogram is misleading under the FIR.

As it is not obvious from ECJ case-law what can be expected from the normative "average consumer" when confronted with visual information and due to the complexity of how consumers process information, legal interpretation may not be conclusive in the doctrinal assessment of whether green pictograms could be

[53] Cf Purnhagen and Van Herpen (2017), n 18, pp. 217–234.

considered misleading under the FIR. This chapter showed that insights from cognitive research and behavioural economics as to how consumers process information may provide further guidance. The use of such insights in the assessment of the normative benchmark of the "average consumer" under the FIR may especially be fitting as the FIR aims at ensuring a high level of consumer protection. It, therefore, seems reasonable to expect that the FIR would allow for behavioural insights to inform legal decision making when applying the normative "average consumer" benchmark to visual information as used by green pictograms. As such, behavioural research may provide the needed clarity as to whether the normative "average consumer" could be considered misled. The current chapter shows that national courts could use these insights to come to a normative decision that is better suited to ensure a high level of consumer protection under the FIR, and, thereby, stimulating sustainable consumption.

References

Cafaggi F, Iamiceli P (2013) Supply chains, contractual governance and certification regimes. Eur J Law Econ 37:131

Cohen J (1988) Statistical power analysis for the behavioral sciences, 2nd edn. Lawrence Erlbaum Associates, Hillsdale

Duivenvoorde B (2015) The consumer benchmarks in the unfair commercial practices directive. Springer Science, pp 29–52

Grunert K, Wills J (2007) A review of European research on consumer response to nutrition information at food labels. J Public Health 15:385

Hahnel U et al (2015) The power of putting a label on it: green labels weigh heavier than contradicting product information for consumers' purchase decisions and post-purchase behavior. Front Psychol 6:1392

Hillman R (2000) The limits of behavioral decision theory in legal analysis: the case of liquidated damages. Cornell Law Rev 85:717

Howells G, Micklitz HW, Wilhelmsson T (2006) European fair trading law – the unfair commercial practices directive. Ashgate Publishing, p 116

Hutchinson T (2015) The doctrinal method: incorporating interdisciplinary methods in reforming the law. Erasmus Law Rev 130

Mak V (2011) Standards of protection: in search of the 'average consumer' of EU law in the proposal for the consumer rights directive. ERPL 15:27–29

Miles S, Ueland O, Frewer L (2005) Public attitudes towards genetically-modified food. Br Food J 107:246

Purnhagen K, Van Herpen E (2017) Can bonus packs deceive consumers? A demonstration of how behavioural consumer research can inform unfair commercial practices law on the example of the ECJ's mars judgment. J Consum Policy 40:217

Rachlinski J (2007) New law and psychology: a reply to critics, skeptics, and cautious supporters. Cornell Law Rev 85:739

Trzaskowski J (2016) Lawful distortion of consumers' economic behaviour – collateral damage under the unfair commercial practices directive. Eur Bus Law Rev 27:33

van Dam YK, van Trijp HCM (2011) Cognitive and motivational structure of sustainability. J Econ Psychol 32:726

Chapter 4
Sustainability Labels in the Shadow of EU Fundamental Rights Law

4.1 Introduction

Practice in the USA shows that some food labelling regulations could be considered inconsistent with the freedom of expression of food businesses.[1] Recent trends in the EU[2] require investigation whether the right to freedom of expression of food businesses could limit the government need to regulate food information.

The objective of this study was to explore to what extent freedom of expression[3] should protect food businesses against government intervention with communications on sustainability labels. A functional comparative method was used to analyse this objective.[4] To achieve this objective, first (1) the different approaches used by the examined legal systems to ensure free speech protection were considered. To do this, it was first considered whether fundamental rights in general and free speech in

---

This chapter is based on Eva van der Zee, 'Legal Limits on Food Labelling Law: Comparative Analysis of the EU and the USA' (2016) 27(3) European Business Law Review.

---

[1] See for example, Case Amestoy *Amestoy*, 92 F.3d 67 (2nd Cir. 1996), Chap. 1, n 58; Case Pearson I *Pearson I*, 164 F.3d 650 (D.C. Cir. 1999), rehearing den., 172 F.3d 72 (D.C. Cir. 1999), Chap. 1, n 58; Case Boggs *Boggs*, 622 F.3d 628 (6th Cir. 2010), Chap. 1, n 58.

[2] In Recital 44 of Regulation (EU) No 609/2013 of the European Parliament and of the Council of 12 June 2013 on food intended for infants and young children, food for special medical purposes, and total diet replacement for weight control it is stated that "[t]his Regulation does not affect the obligation to respect fundamental rights and fundamental legal principles, including the freedom of expression, as enshrined in Article 11, in conjunction with Article 52, of the Charter of Fundamental Rights of the European Union, and in other relevant provisions." Furthermore, Case 544/10 *Weintor*, ECLI:EU:C:2012:526, Chap. 1, n 58 was the first, and so far only, ECJ-case concerning a fundamental rights challenge against a food labelling law.

[3] There seems to be no difference between the term 'freedom of expression' used by civil law systems and 'freedom of speech' used by common law systems. The two terms will, therefore, be used interchangeably throughout this chapter. *See* Barendt (2011), p. 893.

[4] Gordley (2012), Chap. 1, n. 55, 113.

© The Author(s), under exclusive license to Springer Nature Switzerland AG 2022
E. van der Zee, *Sustainability Labels in the Shadow of the Law*, Studies in European Economic Law and Regulation 22, https://doi.org/10.1007/978-3-030-95802-2_4

particular, are protected in the legal system at issue. If this was the case, it was then considered whether and why this protection includes businesses. It was furthermore considered whether this protection is extended to sustainability labels. When freedom of expression applies to communications on sustainability labels it was considered which limitations can be set to such free speech protection. Second (2), the different approaches to free speech protection on food labels in the USA and the EU were compared. Finally (3), it was discussed to what extent the functions of free speech allow free speech protection of communications on sustainability labels.

The analysis is centred on five situations in which government regulation could potentially be considered inconsistent with the freedom of expression:

Cooked-up is a food business producing canned dinners, such as *Mac'n'Cheese* (a canned macaroni and cheese dinner). *Cooked-up* wants to label *Mac'n'Cheese* with the statement that it 'contains all the nutrients needed for a long and healthy life'. The government bans this information as it is considered to be false.

Corn Rebel is a food business producing GM-free sweet corn. According to *Corn Rebel* genetically engineered crops and food products are unsafe for human consumption and hazardous for the environment. *Corn Rebel* wants to disclose on its food label that its sweet corn is 'GM-free' to strengthen the political debate against GM foods. The government prohibits the claim because it finds that the claim confuses consumers, as there is no sufficient scientific evidence that GM products differ compositionally from non-GM products.

True-blue is a food business producing *Blueberrylicious* (blueberry flavoured jelly beans with added Vitamin C). They want to disclose on the food label of *Blueberrylicious* that it contains 'added Vitamin C'. Although the government considers it truthful information, it is nonetheless prohibited because the government found that the statement will contribute to the problem of obesity as it will encourage consumers to eat unhealthy food products.

My Goodness is a food business producing dairy products containing Bifidus. Based on minority scientific opinion *My Goodness* wants to disclose on the food label that 'the consumption of Bifidus eases the digestive system'. The government prohibits the claim because there is no significant scientific evidence supporting the claim.

Humble Honey is a food business producing honey. Their honey might be inadvertently contaminated with genetically modified pollen. The government compels *Humble Honey* to label its honey as being 'contaminated with genetically modified pollen' to enable consumers to make an informed choice. *Humble Honey* does not want to disclose such information as they worry it may negatively affect their sales.

The legal systems of the EU and the USA were subject to the comparison. These cases encompass legal systems from (common and civil) legal cultures at comparable stages of cultural, political and economic development,[5] but have fundamentally different labelling requirements for food products.

The challenges involved with conducting comparative legal research were controlled by extensively consulting experts in constitutional law and food law in each of the two legal systems. This will prevent systematic differences in interpretation and missing out on certain rules and considerations in the foreign system.

[5] I have selected the cases based on the "most similar cases" logic, described in Hirschl (2005), p. 125.

The study was restricted to freedom of expression on the physical label on the food product. All other types of expression concerned with the food product, such as online information, were excluded. Future research may aim at more comprehensive study concerning how online food information is, could, and should be regulated in accordance with freedom of expression.

Government regulations comparable to food labelling regulations were included when these regulations touch upon similar legal dilemmas. Examples of such government regulation include case law concerning regulation prohibiting or limiting advertisement, sponsorship and/or labelling of tobacco products, case law considering labelling of diet supplements, and case law concerning the labelling of alcoholic beverages.[6]

4.2 Free Speech Protection in the European Union

The right to freedom of expression in the EU is enshrined in Article 11 of the Charter of Fundamental Rights of the European Union (hereinafter: CFR). The CFR was proclaimed in December 2000 but did not acquire legally binding status[7] until an amendment of Article 6 TEU in 2009.[8]

---

[6] In the EU, alcoholic beverages are considered to be food products. In the US, however, alcoholic beverages are regulated primarily by the Bureau of Alcohol Tobacco and Firearms, and not by the Food and Drug Authority.

[7] Although the ECJ already cited the CFR before, see e.g. Case 540/03 *Parliament v Council*, ECLI: EU:C:2006:429. Also advocates general already discussed the CFR (See e.g. *TNTTraco*, Case 340/99, Advocate General Alber's opinion 2001, ECR I-4109; *BECTU*, Case 173/99, Advocate General Tizzano's opinion 2001, ECR I-4881; *D and Sweden* v *Council*, Case C-122 & 125/99 P, Advocate General Mischo's opinion 2001, ECR I-4319; *Z* v *Parliament*, Case C-270/99 P, Advocate General Jacobs' opinion 2001, ECR I-9197; *Commission* v *Italy*, Case C-49/00, Advocate General Stix-Hackl's opinion 2001, ECR I-8575; *The Netherlands* v *Council* , Case C-377/98, Advocate General Jacobs' opinion 2001, ECR I-7079; *Council* v *Hautala*, Case C-353/99 P, Advocate General Léger's opinion 2001, ECR I-9565; *Booker Aquaculture Ltd* v *Scottish Ministers*, Case C-20&64/00, Advocate General Mischo's opinion 2003, ECR I-7411; *Überseering*, Case C-208/00, Advocate General Ruiz-Jarabo's opinion 2002, ECR I-9919; *Arben Kaba* v *Secretary of State for the Home Department*, Case C-466/00, Advocate General Ruiz-Jarabo's opinion 2003, ECR I-2219; *Evans*, Case C-63/01, Advocate General Alber's opinion 2003, ECR I-14447; *Omega*, Case C-36/02, Advocate General Stix-Hackl's opinion 2004, ECR I-9609; *Nardone*, Case C-181/03 P, Case C-181/03 P, Advocate General Poiares Maduro's opinion 2005, ECR I-199; *Berlusconi and Others*, Case C-387/02, Advocate General Kokott's opinion 2005, ECR I-3565; *Regione autonoma Friuli-Venezia Giulia and ERSA*, Case C-347/03, Advocate General Jacobs's opinion 2005, ECR I-3785); Furthermore, the CFR gained momentum in secondary law (E.g. Council Directive 2003/86/EC of 22 September 2003 on the right to family reunification [2003] OJ L 251, recital 2; Council Directive 2004/113/EC of 13 December 2004 implementing the principle of equal treatment between men and women in the access to and supply of goods and services OJ L 373, recital 4).

[8] The amended Article 6(1) TEU states that the EU 'recognises the rights, freedoms and principles set out in the Charter of Fundamental Rights of the European Union (...) which shall have the same legal value as the Treaties'.

Article 11 CFR stipulates that:

> Everyone has the right to freedom of expression. This right shall include freedom to hold opinions and to receive and impart information and ideas without interference by public authority and regardless of frontiers.

The meaning and scope of the CFR rights are determined by case law of the European Court of Justice (ECJ), and may also be determined by case law of the European Court of Human Rights (ECtHR)[9] without thereby adversely affecting the autonomy of Union law and of that of the ECJ.[10] Furthermore, in so far as the CFR contains rights that correspond[11] to rights guaranteed by the European Convention on Human Rights (ECHR), the meaning and scope of those rights shall be the same as those laid down by the ECHR.[12] Whether a right has the same meaning and scope to those guaranteed by the ECHR is elaborated upon in the explanations relating to

---

[9]The ECtHR is a supranational or international court established by the European Convention on Human Rights (ECHR). The ECHR is an international treaty, drafted within the Council of Europe, now including 47 members that was formed after the Second World War in an attempt to unify Europe. Ten countries founded the Council of Europe on 5 May 1949: Belgium, Denmark, France, Ireland, Italy, Luxembourg, Norway, Sweden, the United Kingdom and The Netherlands. Today, the Council of Europe covers almost the entire European continent, with its 47 member countries: Albania, Andorra, Armenia, Austria, Azerbaijan, Belgium, Bosnia and Herzegovina, Bulgaria, Croatia, Cyprus, Czech Republic, Denmark, Estonia, Finland, France, Georgia, Germany, Greece, Hungary, Iceland, Ireland, Italy, Latvia, Liechtenstein, Lithuania, Luxembourg, Malta, Republic of Moldova, Monaco, Montenegro, The Netherlands, Norway, Poland, Portugal, Romania, Russian Federation, San Marino, Serbia, Slovak Republic, Slovenia, Spain, Sweden, Switzerland, Macedonia, Turkey, Ukraine and the United Kingdom. The Council of Europe remains entirely independent and separate from the EU, and has no powers in prescribing law to its members. The EU is not a member to the ECHR.

[10]5th recital of the CFR Preamble; Explanations relating to the Charter of Fundamental Rights [2007] OJ C303/17; concerning explanation to Article 52(3), In Opinion Pursuant to Article 218(11) TFEU—Draft International Agreement—Accession of the European Union to the European Convention for the Protection of Human Rights and Fundamental Freedoms—Compatibility of the Draft Agreement with the EU and FEU Treaties, Opinion 2/13 (Opinion of the Full Court, Dec. 18, 2014) restated the autonomy of Union law and the ECJ.

[11]Explanations relating to the Charter of Fundamental Rights, n 10.

[12]Article 52(3) CFR. The reference to the ECHR also includes the Protocols to the ECHR. See, Explanations relating to the Charter of Fundamental Rights, n 10. The CFR also includes a large number of social and economic rights derived from the European Social Charter (ESC). The ESC is a treaty drafted by the Council of Europe in 1961 and it was revised in 1996. The ESC only asks from the State Parties to submit reports indicating how they implement the provisions of the ESC to the European Committee of Social Rights (Article 21 and 22 ECR as amended by the 1991 Turin Protocol). Any decision the ESC takes based on these reports are not binding on the State Parties (Article 28 ECR as amended by the 1991 Turin Protocol; Article 8 and 9 Additional Protocol to the European Social Charter (1995)). The ECtHR has had, due to the binding nature of its judgments, a much greater impact on the rights enshrined in the ECHR, than the ESC has had on the rights enshrined in the ESC. The rights enshrined in the ECHR are, therefore, more developed than the rights enshrined in the ESC.

the CFR.[13] According to the explanations, Article 11 CFR has the same meaning and scope as Article 10 ECHR.[14] Although the explanations do not have the status of law, "they are a valuable tool of interpretation intended to clarify the provisions of the Charter".[15] Furthermore, three references in EU primary law can be found that confirm that the CFR has to be interpreted with due regard to the explanations (Article 52(7) CFR; fifth recital to the preamble of the CFR; Article 6(1) TEU).[16] The ECJ indeed also appears to follow the explanations.[17]

According to the explanations the level of protection afforded by the CFR to rights that have the same meaning and scope to those guaranteed by the ECHR may not be lower than guaranteed by the ECHR.[18] EU law may, however, provide "more extensive protection".[19] This suggests that the CFR interprets the ECHR as a minimum standard of protection, but not necessarily as a maximum.[20]

4.2.1 Scope of Free Speech Protection of Food Businesses in the EU

There is no dispute that companies and corporate entities enjoy fundamental rights protection in the EU.[21] It is not clear, however, whether companies are excluded from fundamental rights protection of strictly personal fundamental rights, such as the right to life (Article 2 CFR) or personal integrity (Article 3 CFR) might be excluded, and arguably freedom of expression.[22] It would be practical if for strictly personal fundamental rights only natural persons could be beneficiaries. The ECJ, however, does not rely on this test.[23]

---

[13] Articles 2, 4, 5(1)+(2), 6, 7, 9, 10(1), 11, 12(1), 14(1)+(3), 17, 19(1)+(2), 47(2)+(3), 48, 49(1) and 50 all correspond to the ECHR or its protocols. See Explanations relating to the Charter of Fundamental Rights, n 10, 17–18.

[14] Ibid., 18.

[15] Ibid.

[16] Weiβ (2012), p. 224.

[17] Case 279/09 *DEB*, ECLI:EU:C:2010:811, at 32, 35-6; Case 283/11 *Sky Österreich*, ECLI:EU:C:2013:28, at 42; Case 334/12 *Réexamen Arango Jaramillo*, ECLI:EU:C:2013:134, at 42; Case 617/10 *Åkerberg Fransson*, ECLI:EU:C:2013:105 at 20.

[18] Explanations relating to the Charter of Fundamental Rights, n 10, 17.

[19] Article 52(3) CFR. This was reconfirmed by CONV 354/02, Final Report of Working Group II, 22 October 2002, p. 7.

[20] Biondi (2012), p. 163.

[21] Case 301/04 P *Commission v SGL Carbon AG*, ECLI:EU:C:2006:53, at 64; Case 249/09 *Novo Nordisk*, ECLI:EU:C:2010:616, at. 44; Case 11/70 *Internationale Handelsgesellschaft*, ECLI:EU:C:1970:114, at 4 ff; Case 136/79 *National Panasonic*, ECLI:EU:C:1980:169 at17 ff; Case 265/87 *Schrader*, ECLI:EU:C:1989:303 at 15.

[22] See also Rengeling and Szczekalla (2004), pp. 344–390; Ehlers (2007), p. 385.

[23] See Ehlers (2007), n. 22, 385–386.

The proceedings of the legitimacy of the Tobacco Advertising Directive[24] in 2000 provided an opportunity for the ECJ to comment on the scope of free speech protection for companies. In this case, Germany sought annulment of the Tobacco Advertisement Directive. The legal challenge raised seven different possible grounds for the annulment of the Directive.[25] One of the grounds was the violation of the right to freedom of commercial expression. The judgment did not address the issue of compatibility with the right to freedom of expression, because the ECJ accepted the lack of a proper legal basis as a ground for annulment of the Directive. Advocate General Fennelly, however, assessed the compatibility of the Directive limiting advertising and sponsorship of tobacco products[26] with the right to freedom of expression. He argues that:

> Personal rights are recognized as being fundamental in character, not merely because of their instrumental, social functions, but also because they are necessary for the autonomy, dignity and personal development of individuals. Thus, individuals' freedom to promote commercial activities derives not only from their right to engage in economic activities and the general commitment, in the Community context, to a market economy based upon free competition, but also from their inherent entitlement as human beings freely to express and receive views on *any* topic, including the merits of the goods or services which they market or purchase.[27]

Although he does not directly address whether companies as such should be beneficiaries of free speech protection, it follows from his submissions that he finds that in this case the companies which manufacture tobacco products are subject to free speech protection, even though he considers it to be a personal right.

The ECtHR elaborated extensively on free speech protection. Whether corporate persons are beneficiaries to the right to freedom of expression as enshrined in Article 10 ECHR was discussed for the first time by the ECtHR in Autronic AG v Switzerland in 1990. This interpretation has become settled case-law of the ECtHR.[28] The ECtHR held that:

> In the Court's view, neither Autronic AG's legal status as a limited company nor the fact that its activities were commercial nor the intrinsic nature of freedom of expression can deprive Autronic AG of the protection of Article 10 (...) The Article (...) applies to 'everyone', whether natural or legal persons.[29]

---

[24] Directive 98/43/EC of the European Parliament and of the Council, issued on 6 July 1998, on the approximation of the laws, regulations, and administrative provisions of the Member States relating to the advertising and sponsorship of tobacco products.

[25] See for more information Weatherill (2011).

[26] Directive 98/43/EC of the European Parliament and of the Council, issued on 6 July 1998, on the approximation of the laws, regulations, and administrative provisions of the Member States relating to the advertising and sponsorship of tobacco products.

[27] Case 376/98 *Germany v Parliament and Council*, ECLI:EU:C:2000:324, at 154.

[28] *Casado Coca v Spain*, app no 15450/89, ECtHR 24 February 1994, 18 EHRR 1, at 35; See Emberland (2006), p. 130.

[29] *Autronic AG v Switzerland*, app no 12726/87, ECtHR 22 May 1990, A/178, 12 EHRR 485, at. 47.

Food businesses would, thus, probably be considered beneficiaries to Article 11 CFR.

4.2.2 Scope of Free Speech Protection of Communications on Food Labels in the EU

Although food businesses most likely will be considered beneficiaries to the right to freedom of expression enshrined in Article 11 CFR, it should still be examined whether communications on the food label could also be considered 'expression' under these articles, especially since in 2013 the EU included freedom of expression in a food labelling regulation.[30]

There is not yet a clear ECJ judgment that would be relevant for expressions on the food label. Resource could be had to the case-law of the ECtHR. 'Expression' in the context of the ECHR is, at least, an expressive statement represented in written or spoken words, pictures, images and expressive conduct, which has an element of public outreach.[31] Besides the expression itself, also the means for its production and its communication, such as print,[32] radio[33] and television broadcasting,[34] artistic creations,[35] film[36] and electronic information systems, is protected.[37] Furthermore, the ECtHR stated in Markt Intern Verlag v Germany that "Article 10(1) (...) does not apply solely to certain types of information or ideas or forms of expression".[38] All[39] forms of expression are, thus, protected by Article 10(1) ECHR.

From this, it could be concluded that corporate communications on the food label, i.e. the statements by *Cooked-up*, *Corn Rebel*, *True-blue*, and *My Goodness* on their

[30] In Recital 44 of Regulation (EU) No 609/2013 of the European Parliament and of the Council of 12 June 2013 on food intended for infants and young children, food for special medical purposes, and total diet replacement for weight control it is stated that "[t]his Regulation does not affect the obligation to respect fundamental rights and fundamental legal principles, including the freedom of expression, as enshrined in Article 11, in conjunction with Article 52, of the Charter of Fundamental Rights of the European Union, and in other relevant provisions."

[31] Emberland (2006), n 28, 117.

[32] *Handyside v UK*, app no 5493/72, ECtHR 7 December 1976, 1 EHRR 737.

[33] *Groppera Radio AG v Switzerland*, app no 10890/84, ECtHR 28 March 1990, 12 EHRR 321.

[34] *Autronic AG v Switzerland*, app no 12726/87, ECtHR 22 May 1990, A/178, 12 EHRR 485.

[35] *Müller and ors v Switzerland*, app no 10737/84, ECtHR 24 May 1988, 13 EHRR 212.

[36] *Otto-Preminger-Institut v Austria*, app no 13470/87, ECtHR 20 September 1994, 17 EHRR.

[37] Harris et al. (1995), pp. 378–379.

[38] *Markt Intern Verlag GmbH and Beermann v Germany*, app no 10572/83, ECtHR 20 November 1989, 12 EHRR 161 at 26.

[39] Hate speech might, however, be excluded from protection. *See* Keane (2007), p. 641 for arguments in favour of excluding hate speech from free speech protection; Cannie and Voorhoof (2011), p. 54 for arguments against excluding hate speech from free speech protection.

Table 4.1 Limitation Clauses in the EU

	Article 52(1) CFR	Article 10(2) ECHR
Content	• provided for by law • respect the essence of those rights and freedoms. • Subject to the principle of proportionality • necessary • genuinely meet objectives of general interest OR need to protect the rights and freedoms of others	• prescribed by law • necessary • legitimate public aim (exhaustively listed)

food labels should thus also be protected by Article 10(1) ECHR, and may, therefore, also be protected by Article 11 CFR.

Whether *Humble Honey*, who is compelled to label its honey as being 'contaminated with genetically modified pollen' to enable consumers to make an informed choice, would enjoy free speech protection in the EU is less clear. The ECJ never discussed such a negative right to freedom of expression. Also, the ECtHR has not explicitly taken a position on whether or not the negative right to freedom of expression is protected by Article 10 ECHR. However, the European Commission on Human Rights (ECmHR)[40] asserted in Goodwin v. United Kingdom that:

> There are circumstances in which a "negative right" is to be implied in Article 10 (Art. 10) not to be compelled to give information or to state an opinion.[41]

Goodwin v. United Kingdom, however, concerned the compulsion of a journalist to disclose its sources. The ECtHR has frequently stressed the importance of the press as a 'public watchdog' to impart information and ideas of public interest.[42] It is, therefore, not self-evident that disclosure requirements on food labels would also enjoy negative free speech protection.

4.2.3 Limitations to the Free Speech Rights of Food Businesses on Food Labels in the EU

Article 52(1) CFR is the overarching limitation clause of the CFR, and closely follows the case-law of the ECJ. When applying Article 52(1) CFR to Article 11 CFR, the explanations relating to the CFR indicate that due regard should be

[40] Initially the ECtHR and the ECmHR where part of the international judicial mechanism with jurisdiction to find against States that breach the rights enshrined in the ECHR. The task of the ECmHR was to screen the incoming cases for admissibility (see former Article 28 ECHR) until it was made defunct in 1998 and its tasks then were taken over by the ECtHR. *See* Emberland (2006) n 28, 9.

[41] *Goodwin v United Kingdom,* app no 17488/90, ECtHR 1 March 1994, 22 EHRR 123, at 48.

[42] Barendt (2011), n 3, 66, refers to *Observer and Guardian v UK* (1992), app no 13585/88, ECtHR 26 November 1991, 14 EHRR 153 at 59; *Jersild v Denmark,* app no 15890/89, ECtHR 23 September 1994, 19 EHRR 1.

given to the limitation clause of the freedom of expression of the ECHR, Article 10(2) ECHR[43] and that government limitations imposed on the right to freedom of expression of Article 11 CFR may "not exceed those provided for in Article 10(2) [ECHR]".[44] Article 10(2) ECHR could, therefore, be taken into account when assessing the limits to free speech protection in the EU. A side-by-side comparison of key phrases from Article 52(1) CFR and Article 10(2) ECHR shows that the articles are very similar. The CFR, however, seems to provide a little more protection than the ECHR (Table 4.1).

So far companies have not challenged a government regulation limiting content on product labels based on the right to freedom of expression before the ECtHR or the ECJ. Nonetheless, the limiting clauses for protection following ECJ case-law and Article 52(1) CFR could play out as follows.

(1) Limitation Must Be Provided for by Law
In cases where it involves food labelling law, this condition is met by definition. Within the European Union many legally binding rules, mainly Regulations but also Directives,[45] relate to the food label.[46] This requirement is similarly stipulated in Article 10(2) ECHR which specifies that the government interference must be prescribed by law, meaning that, at a minimum, the interference should be authorized by a specific national, European or international legal rule or regime.[47]

(2) Limitation Must Respect the Essence of the Rights and Freedoms at Issue
Case-law of the ECJ indicates that interferences with the fundamental rights of the EU may not impair the very essence of those rights.[48] The wording of Article 52(1) CFR is based on the case-law of the ECJ, which holds that "restrictions may be imposed on the exercise of fundamental rights (...) provided that those restrictions (...) do not constitute, with regard to the aim pursued, disproportionate and unreasonable interference undermining the very *substance* of those rights".[49] Adherence to the essence of a fundamental right, however, "does not require more than the

[43] The rights and freedoms enshrined in the ECHR each have their own limitation clause, instead of one overarching limitation clause as in the CFR.

[44] Explanations relating to the Charter of Fundamental Rights, n 10, at 1.

[45] Regulations are defined in the Treaty on the Functioning of the European Union as having general application and 'binding in its entirety and directly applicable in Member States' (Article 288 TEU).

[46] Examples are EU Food Information Regulation, Chap. 1, n 57; EC Regulation on Organic Agriculture, Chap. 2, n 40; Regulation (EC) 1829/2003 on genetically modified food and feed, OJ L 268; Regulation (EC) 1830/2003 concerning the traceability and labelling of genetically modified organisms and the traceability of food and feed products produced from genetically modified organisms [2003] OJ L 268; Regulation (EC) 1924/2006 on the nutrition and health claims made on foods; Regulation (EC) 1760/2000 establishing a system for the identification and registration of bovine animals and regarding the labelling of beef and beef products [2000] OJ L 204.

[47] Harris et al. (1995), n 37, 345; *Silver v UK*, app no 5947/72, ECtHR 25 March 1983, at 86.

[48] Case 5/88 *Wachauf*, ECLI:EU:C:1989:321, at 18.

[49] Case 292/97 *Karlsson and Others*, ECLI:EU:C:2000:202, at 53.

preservation of all basic guarantees which emanate from the right in question".[50] It could be argued, therefore, that some food labelling regulations affect the essence of the right to freedom of expression (Article 11 CFR) by prohibiting a form of expression. The exact essence of free speech according to the ECJ is, however, yet unclear.

(3) Subject to the Principle of Proportionality

The principle of proportionality requires that measures adopted by EU institutions do not exceed the limits of what is appropriate and necessary to attain the objectives legitimately pursued by the legislation in question. As such the limitation should be necessary (see 3.1. below), and genuinely meet objectives of general interests or the need to protect the rights and freedom of others (see 3.2. below).

(3.1) The Limitation Must Be Necessary

According to the ECJ, the necessity-requirement implies that the limitation should be the least onerous option of the available options.[51] The ECJ had not yet elaborated on this requirement, although Advocate General Fennelly argues that when an EU measure restricts freedom of commercial expression the EU legislator should:

> be obliged to satisfy the Court that it had reasonable grounds for adopting the measure in question in the public interest. In concrete terms, it should supply coherent evidence that the measure will be effective in achieving the public interest objective invoked (. . .) and that less restrictive measures would not have been equally effective.[52]

Advocate General Jääskinen argued that in commercial matters the EU legislator has wide discretion in assessing the level of public health protection and is not required to restrict itself to a minimum necessity to protect freedom of expression.[53] Fennelly adds nuance by stating that "[t]he more restrictive the effects, the greater is the onus on the legislator to show that a less burdensome measure would not have sufficed". Fennelly suggests that the "evidence required to justify a restriction will depend on the nature of the claim made,"[54] because "[e]vidづentiary requirements may be less strict where public health is at stake",[55] implying that public health by definition gives strong support for any type of restriction. This is in line with the ECtHR which held in two cases concerning tobacco advertising that "overriding considerations of public health, on which the State and the European Union have,

[50] Ehlers (2007) n 22, 393.

[51] See Case 283/11 *Sky Österreich*, ECLI:EU:C:2013:28, n 17, at 50.

[52] Case 376/98 *Germany v Parliament and Council*, ECLI:EU:C:2000:324, n 27, at 159.

[53] Case 249/09 *Novo Nordisk*, ECLI:EU:C:2010:616, n 21, at.50. He refers to Alexy (2003), p. 433, 440.

[54] Case 376/98 *Germany v Parliament and Council*, ECLI:EU:C:2000:324, n 27, at 160.

[55] Ibid, n 27, at 161.

moreover, legislated, may take precedence over economic concerns, and even over certain fundamental rights such as freedom of expression".[56]

Limiting false commercial speech, such as *Cooked-up*'s false claim that its canned macaroni and cheese dinner 'contains all the nutrients needed for a long and healthy life', will most likely be considered necessary. It may be different for *Corn Rebel*'s claim that its sweet corn is being 'GM-free', *True-blue*'s claim that Vitamin C is added to its *Blueberrylicious* treats, or *My Goodness*'s claim that 'the consumption of Bifidus eases the digestive system'. It could well be that adding disclaimers (for example for *Corn Rebel*: 'No significant difference has been shown GM corn and non-GM corn'; for *True-blue*: 'Although the treats contain Vitamin C they should still be eaten in moderation'; and for *My Goodness*: 'This claim is based on minority scientific evidence') would be preferred, being the least onerous option of the available options, as long as it is equally effective as banning the information altogether.

The ECJ did already prefer disclaimers over a prohibition with respect to the free movement of goods. In the Cassis de Dijon ruling[57] and the Beer Purity-case[58] the governments invoked consumer protection to restrict the trade of certain products.[59] The ECJ found that disclaimers to the product in question were preferred because they were less restrictive to trade and had the same effectiveness as prohibiting trade of the product altogether.[60] The ECJ would likely adopt a similar approach when the fundamental right to freedom of expression is limited.

The necessity requirement can also be found in Article 10(2) ECHR which holds that any interference to freedom of expression must be 'necessary in a democratic society'. According to the ECtHR interference would be 'necessary in a democratic society' when the interference corresponds to a pressing social need and that the interference is proportionate to the legitimate aim pursued.[61] Furthermore, the reasons for the interference must be relevant and sufficient. In assessing whether and to what extent government interference is necessary the governmental authorities have the 'margin of appreciation'.[62] The margin of appreciation is not unlimited and could even be reduced to zero. The ECtHR can give a final ruling on whether government interference is reconcilable with freedom of expression.[63] Relevant for

---

[56] Translation from Case 249/09 *Novo Nordisk*, ECLI:EU:C:2010:616, n 21, at 46. The cases are only accessible in French. See *Société de Conception de Presse et d'Edition et Ponson v France*, no 26935/05, ECtHR 6 March 2009; and *Hachette Filipacchi Presse Automobile and Dupuy v France*, app no 13353/05, ECtHR 5 March 2009: "Ainsi, des considérations primordiales de santé publique, sur lesquelles l'Etat et l'Union européenne ont d'ailleurs légiféré, peuvent primer sur des impératifs économiques, et même certains droits fondamentaux comme la liberté d'expression."

[57] Case 120/78 *Cassis de Dijon*, ECLI:EU:C:1979:42.

[58] Case 178/84 *Beer Purity*, ECLI:EU:C:1987:126.

[59] Ibid.

[60] See further on this information paradigm Purnhagen (2014), pp. 329–332.

[61] *Olsson and Olsson v Sweden*, app no 10465/83, ECtHR 24 March 1988, 11 EHRR 259 at 67 PC.

[62] *Handyside v UK*, app no 5493/72, ECtHR 7 December 1976, 1 EHRR 737, at 48-9 PC.

[63] *Handyside v UK*, app no 5493/72, ECtHR 7 December 1976, 1 EHRR 737, at. 49.

this study is that the width of the margin of appreciation is wider when the expression is considered to be commercial in nature rather than political. Thus, when the statements by *Corn Rebel* are considered political in nature, the width of the margin of appreciation is wider than the overtly commercial statements by *Cooked-up*, *True-blue*, and *My Goodness.*

The ECtHR defines commercial expression as "inciting the public to purchase a particular product".[64] Commercial expression is aimed at enhancing the economic interests of individuals and businesses.[65] According to the ECtHR political expression concerns the speaker's "participation in a debate affecting the general interest"[66] or reflects "controversial opinions pertaining to modern society in general".[67] For example, expression that is considered to contribute to public debate, even when it boosts the businesses of the speaker, should not be classified as commercial expression.[68]

The ECtHR, furthermore, made a distinction between 'pure' commercial expression and commercial expression with 'political overtones'[69] 'Purely' commercial expression has no political overtones and is subject to the lenient *Markt Intern* standard.[70] The *Markt Intern* standard implies that the ECtHR must "confine its review to the question whether the measures taken on the national level are justifiable in principle and proportionate".[71] To establish whether such interference would be proportionate the ECtHR must "weigh the requirements of the protection of the reputation and the rights of others against the publication of the information".[72] When commercial expression concerns an ongoing political debate commercial expression may be considered to have 'political overtones'.[73] Such was the case in Hertel V Switzerland where the appropriateness under Article 10(2) ECHR of court-sanctioned injunctions sought by an association of manufactures against the applicant, who had violated domestic competition laws by publishing statements of the alleged hazards involved in the use of microwave ovens, was considered by the ECtHR. Since the statements concerned an ongoing debate of the effects of microwaves on human health, the Hertel claim was "substantially different from... markt intern" and it was therefore "necessary to reduce the extent of the margin of application" implied in that judgment.

---

[64] *Verein gegen Tierfabrieken v Switzerland*, app no 24699/94, ECtHR 28 September 2001.

[65] Harris et al. (1995) n 37, 461.

[66] *Hertel v Switzerland Reports,* app no 25181/94, ECtHR 25 August 1999, 28 EHRR 534, at 47.

[67] *Verein gegen Tierfabrieken v Switzerland*, app no 24699/94, ECtHR 28 September 2001 at 70.

[68] *Barthold v Germany,* app no 8734/79, ECtHR 25 March 1985, 7 EHRR 383.

[69] See also Emberland (2006) n 28, 165 and 171.

[70] Ibid 164–171.

[71] *Markt Intern Verlag GmbH and Beermann v Germany*, app no 10572/83, ECtHR 20 November 1989, 12 EHRR 161 at 33.

[72] Ibid at 34.

[73] *Hertel v Switzerland Reports,* app no 25181/94, ECtHR 25 August 1999, 28 EHRR 534 at 47.

Commercial expression with 'political overtones' will be subject to a more rigorous scrutiny:[74] the ECtHR could also review whether the interference corresponds to a pressing social need and whether the interference is proportionate to the legitimate aim pursued.[75]

Thus, the claims put forward by *Cooked-up*, *True-blue*, and *My Goodness* may be considered purely commercial. Also, *Corn Rebel*'s claim that its sweet corn is being 'GM-free' should be considered purely commercial, because *Corn Rebel* does not participate in a political debate by labelling their corn as 'GM-free' as such label does not make a statement about the hazards of GM-food for human health or the environment.

(3.2a) The Limitation Must Genuinely Meet Either Objectives of General Interest Recognized by the EU

In settled case-law of the ECJ fundamental rights of the EU may only be restricted for reasons that correspond to "objectives of general interest pursued by the Community".[76] Arguably these general interests are similar to the general interests in the field of free movement of goods, people, services and capital, which include the written grounds of Article 36, 45(4), 52, 62, and 65 TFEU (public morality, policy, or security; protection of health and life of humans, animals, or plants; protection of national treasures possessing artistic, historic or archaeological value; protection of industrial and commercial property) and unwritten grounds in the public interest which are determined in the case-law of the ECJ, including, amongst others, protection of public health,[77] the defence of the consumer,[78] and protection of the environment.[79] Furthermore, according to settled ECJ case law purely economic objectives cannot constitute an overriding reason in the public interest.[80] The ECJ has held that the protection of health is an objective of general interest that follows from Article 9 TFEU.[81] Other objectives of general interest may include Article 7 to 12 TFEU (consistency between policies; eliminate inequalities/promote equality between men and women; promotion of high level of employment; guaranteeing adequate social protection; fight against social exclusion; promotion of high level of education and training' protection of human health; combatting discrimination based on sex, racial or ethnic origin, religion or belief, disability, age or sexual orientation; promoting sustainable development; protecting consumers).

[74] Emberland (2006) n 28, 170.

[75] *Olsson and Olsson v Sweden*, app no 10465/83, ECtHR 24 March 1988, 11 EHRR 259 at 67 PC.

[76] Case 5/88 *Wachauf*, ECLI:EU:C:1989:321, n 48, at 18; Case 292/97 *Karlsson and Others*, ECLI:EU:C:2000:202, n 49, at 45.

[77] Case 544/10 *Weintor*, ECLI:EU:C:2012:526, Chap. 1, n 58.

[78] Case 120/78 *Cassis de Dijon*, ECLI:EU:C:1979:42, n 57, at 8.

[79] Case 142/05 *Aklagaren v Mickelsson and Roos*, ECLI:EU:C:2009:336.

[80] See e.g. Case 436/00 *X and Y*, ECLI:EU:C:2002:704 at 50; Case 35/98 *Verkooijen*, ECLI:EU:C:2000:294. See Trstenjak and Beysen (2013), p. 293, footnote 40.

[81] Article 9 TFEU: "In defining and implementing its policies and activities, the Union shall take into account requirements linked to the (. . .) protection of human health."

(3.2b) Or the Limitation Protects the Rights and Freedoms of Others
A food labelling regulation could also limit free speech to protect the rights of others, which could potentially include the right to health protection or the right to receive information. The ECJ seems to recognize a fundamental right to health protection from the second sentence of Article 35 CFR, which requires that "a high level of human health protection shall be ensured in the definition and implementation of all the Union's policies and activities".[82] Advocate General Jääskinen, who assessed for the first time so far commercial expression in the context of Article 11 CFR, also derived a fundamental right to health protection from Article 35 CFR.[83] He further argued that this fundamental right to health protection must be safeguarded to guarantee the fundamental rights, human dignity, the right to life and the right to physical and mental integrity.[84] He argued that the right to life, and as such the protection of health, must take precedence over the fundamental right to freedom of action, such as the freedom of expression.[85]

Another fundamental right that may justify a limitation to free speech is the right to receive information as stipulated in Article 11 CFR. This right is especially important when free speech protection would also include the right not to speak. However, negative expression on food labels (such as *Humble Honey*'s refusal to disclose that its honey is 'contaminated with genetically modified pollen') will most likely not enjoy free speech protection in the EU (see paragraph 2.2). However, if free speech rights would include the right not to speak, it is not clear whether this right to receive information could confer rights on consumers to demand disclosure of information on the food label. Article 169 TFEU recognizes that consumers have a *right* to information.[86] This treaty provision does, however, not confer rights on consumers as such but it imposes an obligation on EU bodies to ensure a high level of consumer protection.[87] Consumer protection, however, does seem to be acknowledged as a fundamental right of consumers in the EU, because Article 38 CFR stipulates that 'Union policies must ensure a high level of consumer protection'.[88] However, Ehlers argues that besides the right of access to data "within the field of the Union's fundamental rights, no room should be given to further increase the subjectivity of the idea of transparency."[89]

[82] Case 544/10 *Weintor*, ECLI:EU:C:2012:526, Chap. 1, n 58, at 47.

[83] Case 249/09 *Novo Nordisk*, ECLI:EU:C:2010:616, n 21, footnote 21.

[84] Ibid., at 49.

[85] Ibid., at 50. He refers to Alexy (2003), n 53, 440.

[86] Article 169 TFEU: "In order to promote the interests of consumers and to ensure a high level of consumer protection, the Union shall contribute to protecting the health, safety and economic interests of consumers, as well as to promoting their *right to information*, education and to organise themselves in order to safeguard their interests."

[87] Devenney and Kenny (2012), pp. 349–350.

[88] Ibid n 87, 350.

[89] Schorkopf (2007), p. 425.

As a whole, the principle of proportionality seems similar to the requirement of Article 10(2) ECHR that the interference to freedom of expression must serve a legitimate aim. Unlike Article 52(1) CFR, which does not exhaustively list the applicable general interests, Article 10(2) ECHR indicates what constitutes a legitimate aim, i.e. the interference must be in the interests of national security, territorial integrity or public safety, for the prevention of disorder or crime, for the protection of health or morals, for the protection of the reputation or the rights of others, for preventing the disclosure of information received in confidence, or for maintaining the authority and impartiality of the judiciary.[90]

According to the explanations limitations to the right to freedom of expression allowed by Article 52(1) CFR may not exceed those provided for in Article 10(2) ECHR.[91] Government regulations that interfere with the freedom of expression must, therefore, at least meet one of the legitimate aims stipulated in Article 10(2) ECHR.[92] Most government interferences could probably be placed under one of the legitimate aims mentioned in Article 10(2) ECHR as the grounds for interference are broad.[93] It is, therefore, likely that when a food labelling regulation aims at protecting health it could be considered to be a legitimate aim under the ECHR.

4.2.4 Conclusion

In theory, the fundamental right to freedom of expression seems open to food businesses and applies to all types of expression, arguably including expression on food labels. In practice, the protection of the fundamental rights of food businesses for communications on the food label to be limited.

Government limitations to *Cooked-up* utterly false claim, that its canned macaroni and cheese dinners are healthy, will likely be considered necessary. It will be more difficult for the government to justify limitations to *Corn Rebel*'s, *True-blue*'s or *My Goodness*'s claim. In these cases, disclaimers may be preferred as long as the disclaimer has the same effectiveness as a complete ban of the information. Limitations should, however, be placed under one of the legitimate aims stipulated in Article 10(2) ECHR. Limitations based on health protection could probably easily be placed under Article 10(2) ECHR. This will, however, be more difficult when the limitation is based on protecting the consumer right to information. However, it seems that such a right will not be relevant in the context of free speech in the EU, as it is most likely that businesses that refuses to disclose government mandated information on its label, such as *Humble Honey*'s refusal to label its honey as

---

[90] Article 10(2) ECHR.

[91] Explanations relating to the Charter of Fundamental Rights, n 10, at 1.

[92] Ibid.

[93] Harris et al. (1995), n 37, 348.

'contaminated with genetically modified pollen', will not enjoy free speech protection.

4.3 Free Speech Protection in the United States of America

The First Amendment states that "Congress shall make no law . . . abridging the freedom of speech . . ."[94] Corporate entities are afforded, subject to the limits discussed below, First Amendment protection.[95]

4.3.1 *Scope of Free Speech of Communications on Food Labels in the USA*

The degree to which content on food labels may enjoy First Amendment protection depends on whether the speech can be categorized as commercial speech because commercial speech receives limited protection.[96]

Commercial speech was carved out by the Supreme Court in 1976 in *Virginia State Pharmacy Board v.* Virginia *Citizens Consumer Council,*[97] and in 1980 in *Central Hudson Gas & Electric Corp. v. Public Service Commission.*[98] The Court defined commercial speech as "speech with does no more than propose a commercial transaction"[99] or is "related solely to the economic interest of the speaker and its audience."[100] This implies that both parties should have an economic interest in the speech,[101] which excludes books, newspapers, and magazines, read for its political, literary, or other public interest content, from the scope of commercial expression.[102]

[94]U.S. Const., Amend. 1 (1791).

[95]*First Nat'l Bank of Boston v Bellotti*, 435 U.S. 765, 778 (1978).

[96]*Central Hudson Gas & Electric v Public Service Comm'n of New York*, 447 U.S. 557, 561 (1980).

[97]*Virginia State Bd. of Pharmacy v Virginia Citizens Consumer Council, Inc.*, 425 U.S. 748, 762 (1976).

[98]*Central Hudson Gas & Electric v Public Service Comm'n of New York*, 447 U.S. 557 (1980). It is interesting to note that in 1942, the *Court in Valentine v Chrestenen*, 316 U.S. 52 (1942), held that commercial speech was not protected by the First Amendment. "We are equally clear that the Constitution imposes no such restraint on government as respects purely commercial advertising." *Valentine*, 316 U.S. at 54.

[99]*Virginia State Bd. of Pharmacy v Virginia Citizens Consumer Council*, Inc., 425 U.S. 748, 762 (1976).

[100]*Central Hudson Gas & Electric v Public Service Comm'n of New York*, 447 U.S. 557, 561 (1980).

[101]Barendt (2011), n 3, 396.

[102]Ibid.

The First Amendment also restricts the ability of the government to compel individuals to engage in certain expressive activities, as such free speech protection also includes the right not to speak.[103] With respect to disclosure requirements in the realm of commercial speech, such as *Humble Honey*'s compelled claim that its honey 'contaminated with genetically modified pollen', the Supreme Court made clear that such negative commercial speech does also enjoy First Amendment protection.[104]

Most importantly, the commercial expression relates *solely* to economic interests. As such, *Cooked-up*'s false claim that its canned macaroni and cheese dinner 'contains all the nutrients needed for a long and healthy life', *True-blue*'s claim that Vitamin C is added to its *Blueberrylicious* treats, *My Goodness* claim that 'the consumption of Bifidus eases the digestive system', and *Humble Honey*'s refusal to label its honey as being 'contaminated with genetically modified pollen' will thus likely be considered commercial speech.

This may be different for *Corn Rebel*'s political claim that its sweet corn is being 'GM-free'[105] as such information may not be provided solely for economic reasons, but takes a line on political questions or makes a contribution to the formation of public opinion.[106] It seems, however, that the courts may not want to take it that far. A Circuit Court of Appeals held that press releases by the National Commission on Egg Nutrition, a producers' consortium, on a matter of current controversy, that there was no scientific evidence that egg consumption increased heart diseases, were considered commercial speech,[107] even though it was not clear whether these press releases related solely to the economic interest.[108] Furthermore, in *Nike, Inc. v. Kasky*[109] the question was whether Nike's response, in the form of press releases and letters to newspapers, university presidents, and athletics directors regarding allegations that the company was mistreating and underpaying workers outside the USA could be classified as commercial speech. The Supreme Court of California did categorize the speech as commercial and, therefore, the response would not enjoy First Amendment protection if found false or misleading. The majority of the Supreme Court, however, held that the case was not yet ripe for full consideration. In his dissent, Judge Breyer argued that the responses were in

[103] *West Virginia Bd. of Education v Barnette*, 319 U.S. 624, 642 (1943); Wooley v Maynard, 430 U.S. 795 (1977).

[104] *Zauderer v Office of Disciplinary Counsel of the Supreme Court of Ohio*, 471 U.S. 626 (1985).

[105] Another issue with 'GM-free' claims on corn is that it is difficult to guarantee that there would be no contamination with GM corns. If there is contamination the claim of GM-free would be false and thus not protected unless political speech.

[106] Barendt (2011), n 3, 396–397. Barendt refers to the *German Supreme Constitutional Court,* 102 BverfGE 347, 359-60 (2001) who held that the civil courts were wrong to interpret Benetton pictorial advertisements protesting against environmental damage, the employment of children, and the spread of AIDS as solely intended to promote the company's economic interest.

[107] *FTC v National Commission on Egg Nutrition,* 517 F 2d 485.

[108] See also Barendt (2011), n 3, 397.

[109] *Nike Inc. v Marc Kasky*, 123 S Ct 2554 (2003).

form and content public, rather than commercial speech, because the responses by Nike were not made in an advertising format, did not propose sales, and concerned an important matter of public controversy—the criticism of its employment practices.[110] He added that the form and content distinguish the speech at issue from purely commercial speech such as "speech—say, the words "dolphin-safe tuna"—that commonly appears in more traditional advertising or labelling contexts." In a Petitioners' brief, it was also argued that

> If the asserted tie-in between a state's regulatory power and the moral conclusions of consumers ever suffices to convert discussion of public issues into lesser protected "commercial speech," that can only be in the context of direct product advertising and product labels, which are least likely to generate reasoned discussion and which are targeted at consumers and affect purchasing decisions in the first instance and shape broader moral judgments only secondarily.[111]

Although the Supreme Court has not yet decided on the issue of whether commercial expression needs to relate *solely* to the economic interest, it is likely that *Corn Rebel's* claim that their sweet corn is being 'GM-free', will be categorized as commercial speech.

4.3.2 Limitations to the Free Speech Rights of Food Businesses on Food Labels in the USA

To determine whether commercial speech would enjoy First Amendment protection, the Supreme Court articulated in *Central Hudson*, a four-part test. First (1), the speech must concern lawful activity and not be misleading.[112] The Court has long held that expression likely to deceive[113] or related to illegal activity[114] is not protected speech. Second (2), the Court will inquire whether the government has asserted a substantial interest in regulating the commercial speech at issue.[115] Common examples of a substantial government interest include preventing consumer confusion,[116] protecting national security,[117] life, health and safety.[118] The government bears the burden to "demonstrate that the harms it recites are real and

[110] *Ibid. See also Barendt (*2011*)*, n 3, 398.

[111] Brief for the petitioners, No. 02-575, at 36.

[112] Ibid, at 566.

[113] *Friedman v Rogers*, 440 U.S. 1, 15-16 (1979).

[114] *Pittsburgh Press Co. v Human Relations Comm'n*, 413 U.S. 376, 388 (1973).

[115] *Central Hudson Gas & Electric v Public Service Comm'n of New York*, 447 U.S. 557, 566 (1980).

[116] Ibid.

[117] *Haig v Agee*, 453 U.S. 280 (1981).

[118] *See* U.S. Const. preamble & amend *V See also*, Gottlieb (1988), p. 917, 948 (discussing compelling government interest in life, health and safety).

that its restriction will in fact alleviate them to a material degree".[119] However, the Supreme Court held that when the government restricts truthful, non-misleading commercial speech for reasons "unrelated to the preservation of a fair bargaining process" strict scrutiny should apply.[120] Although strict scrutiny is almost always fatal to the challenged government restriction, a Court will uphold the constraint on speech if it is "necessary, and narrowly drawn, to serve a compelling state interest."[121]

If the answer to the first two questions is yes, the Court will then (3) determine "whether the regulation directly advances the governmental interest asserted"[122] and (4) "whether it is not more extensive than necessary to serve that interest."[123] The government needs to establish that the regulation on speech is "narrowly tailored to achieve the desired objective."[124] The government may prohibit inherently misleading advertising, but cannot place an absolute prohibition on *potentially* misleading information, if the information may be presented in a way, such as the use of a disclaimer, that is not deceptive.[125]

Whether information may be considered potentially misleading and, therefore, require a disclaimer was discussed by Court of Appeals with respect to health claims[126] that have some scientific support, such as *My Goodness*'s claim that 'the consumption of Bifidus eases the digestive system', but do not satisfy the Food and Drug Administration's (FDA) "significant scientific agreement" standard.[127] In *Pearson I*, a Court of Appeals noted that in cases of incomplete advertising, the message is not inherently misleading (and thus properly restricted) but rather potentially misleading and that the preferred remedy is more disclosure rather than an outright prohibition.[128] The Court of Appeals held, however, that a disclaimer would not have been necessary when (1) evidence in support of the claim is qualitatively weaker than evidence against the claim; or (2) evidence in support of the claim is outweighed by evidence against the claim.[129] In *Pearson II*, a Court of Appeals added that although significant evidence in support of the claim was absent,

[119] *Edenfield v Fane*, 507 U.S. 761 (1993).

[120] *44 Liquormart, Inc. v Rhode Island*, 17 U.S. 484 (1996).

[121] *Capitol Square Review and Advisory Bd. v Pinette*, 515 U.S. 753, 761 (1995); *Perry Ed. Ass'n v Perry Local Educators' Ass'n*, 460 U.S. 37, 45 (1983); *Carey v Brown*, 447 U.S. 455, 461 (1980).

[122] *Central Hudson Gas & Electric v Public Service Comm'n of New York*, 447 U.S. 557, 566 (1980).

[123] Ibid.

[124] *Board of Trustees v Fox*, 492 U.S. 469, 480 (1989).

[125] *In re R.M.J.*, 455 U.S. 191, 203 (1982).

[126] Health claims describe a relationship between a nutrient, such as calcium, and a disease or health-related condition, such as osteoporosis. 21 U.S.C. § 343(r)(1)(B).

[127] The FDA implemented a rule that required "significant scientific agreement" regarding the link between the claimed nutrient and health impact before allowing use of the claim on a food or supplement label. *See* 21 C.F.R. § 101.14; 21 C.F.R. § 101.70.

[128] *Pearson v Shalala*, 164 F.3d 650, 657 (D.C. Cir. 1999).

[129] Ibid.

this does not mean that it is negative evidence against the claim.[130] The Court of Appeals added in *Pearson II* that disclaimers are not necessary when the government demonstrates "with empirical evidence that disclaimers would bewilder consumers and fail to correct for deceptiveness".[131] In *Whitaker*, the Court of Appeals found that health claims on dietary supplements considering *treating* a disease instead of *reducing* disease *risk* are unlawful, and therefore fail the first part of the *Central Hudson* test.[132]

Thus, when food businesses want to put content on their food labels, such as *Cooked-up*, *Corn Rebel*, *True-blue*, and *My Goodness*, the *Central Hudson* test most likely would apply. This may be different for *Humble Honey*'s refusal to label its honey as being 'contaminated with genetically modified pollen'. Although the government may require food businesses to place labels on their products to regulate commerce[133] or protect the liberty interests of other members of society[134] some examples of challenges to government compelled speech through disclaimers exist mainly in the biotechnology context (*Amestoy* case and *Boggs* case) and tobacco warning labels (*Discount Tobacco* case and *R.J. Reynolds* case).

In the *Amestoy* case, a Court of Appeals invalidated Vermont's mandatory disclosure requirements for dairy products derived from cows treated with a genetically engineered version of bovine somatotropin,[135] commonly referred to as rBST.[136] The Court of Appeals applied the *Central Hudson* test and held that Vermont has failed to establish that its interests are substantial.[137] The Court held that the dairy producers and retailers had a First Amendment right not to speak unless the state could establish a substantial interest in labelling rBST derived products.[138] Vermont argued that its statute supported a "strong consumer interest

[130] *Pearson v Shalala*, 130 F. Supp. 2d 105 (D.D.C. 2001).

[131] *Pearson v Thompson*, 141 F. Supp. 2d 105 (D.D.C. 2001).

[132] *Whitaker*, 239 F. Supp. 2d (D.C. 2003) at 54.

[133] *See e.g.*, U.S. *v* 40 Cases, More or Less, Pinocchio Brand 75% Corn, Peanut Oil and Soya Bean Oil Blended with 25% Pure Olive Oil, 289 F.2d 343, 345 (2d Cir. 1961) ("The interest of the federal government in ensuring that such food meets minimum standards of purity and is not misbranded arises out of its supervisory function over interstate commerce.").

[134] *Jacobson v Massachusetts*, 197 U.S. 11, 26 (1905).

[135] Recombinant Bovine Somatotropin, also known as recombinant Bovine Growth Hormone (rGBH), is a synthetic growth hormone that increases milk production by cows.

[136] Vt. Stat. Ann. tit. 6, § 2754 (terminated by 1993, Adj. Sess., No. 127, § 4, as amended by 1997, No. 61 § 272i, eff. Mar. 30, 1998).

[137] Case Amestoy *Amestoy*, 92 F.3d 67 (2nd Cir. 1996), Chap. 1, n 58; at 73.

[138] Ibid., at 71.

and the public's 'right to know'."[139] The Court, however, held that a "substantial state interest" cannot be established based merely on consumer curiosity.[140]

In *Boggs*, the Court of Appeals did not invalidate Ohio's mandatory disclosure requirements; albeit that this time the mandatory disclosure requirements considered dairy products derived from cows *not* treated with rBST. In Ohio such products should be accompanied by a disclaimer stating that "The FDA had determined that no significant difference has been shown between milk derived from rBST-supplemented and non-rBST-supplemented cows".[141] The Court of Appeals used the *Zauderer* test to assess whether the rule conflicted with the First Amendment.[142] In *Zauderer*, the Supreme Court expressed a lighter standard than the Central Hudson test, applying only to disclosure requirements. In *Zauderer* the Supreme Court held that the government may compel disclosure requirements associated with product marketing, so long as the disclosure is (1) purely factual and uncontroversial; (2) reasonably related to the State's interest in preventing deception of consumers; and (3) not unjustified or unduly burdensome.[143] Accordingly, in *Boggs* the Court concluded that the use of a disclaimer accompanying the production claim could eliminate any consumer confusion and was, therefore, considered not to violate the First Amendment even though it compelled food businesses to speak.[144]

Cases in the context of tobacco warning labels illustrate the difficulty in determining whether the compelled commercial speech at issue is purely factual and uncontroversial. The *Discount Tobacco* case concerned labelling restrictions on tobacco products—specifically the use of colour graphics depicting the negative health consequences of smoking along with textual warning labels.[145] The Court of Appeals for the Sixth Circuit also based its decision on *Zauderer* and upheld the graphic-warning requirement because the factual information (i.e., colour graphics)

---

[139] Ibid., at 73.

[140] Ibid. The *Amestoy* opinion included a vigorous dissent asserting that the state interest was not limited to consumer curiosity, but also substantive concerns regarding rBST's impact on the heath of humans and cows, the financial sustainability of small farms, and general concerns regarding the manipulation of nature using biotechnology. *Id.* at 74. The proper question, in the dissent's view, is whether the Constitution prohibits government from mandating disclosure of truthful, relevant information to promote informed consumer choice. *Id.* Although the Second Circuit opinion certainly leaves open the possibility that mandatory labelling could pass constitutional muster if the state advanced a more substantive interest, a generalized interest in satisfying consumer curiosity appears to be a losing argument for states attempting to mandate labelling of otherwise scientifically indistinguishable products. Rather, the court relegated process-based labelling decisions to market forces.

[141] 60 Ohio Admin. Code § 901:11-8 (2008).

[142] *Zauderer v Office of Disciplinary Counsel of the Supreme Court of Ohio*, 471 U.S. 626 (1985).

[143] Ibid.

[144] Case Boggs *Boggs*, 622 F.3d 628 (6th Cir. 2010), Chap. 1, n 58.

[145] *See Discount Tobacco City & Lottery, Inc. v United States*, 674 F.3d 509 (6th Cir. 2012); R.J. *Reynolds Tobacco Co. v Food and Drug Administration*, 696 F.3d 1205 (D.C. Cir. 2012).

regarding the health risks of using tobacco are reasonably related to the alleviation of potential consumer confusion.[146] In contrast, the Second Circuit Court of Appeals in *R.J. Reynolds*, held that the graphic warnings required under the Act went beyond a full disclosure requirement as in *Zauderer*, to prevent consumer deception, but rather required a general disclosure about the negative health effects of smoking—thus amounting to a warning and discouragement to consumers to purchase products rather than rectify specific deceptive statements.[147] Accordingly, the Court applied the more restrictive *Central Hudson* test—finding that the government failed to present any evidence that the proposed graphics would accomplish the stated goal of reducing smoking rates.

Whether the rational basis test outlined in *Zauderer* and applied in *Discount Tobacco* and *Boggs*, would apply to *Humble Honey*'s refusal to label its honey as being 'contaminated with genetically modified pollen', or whether the intermediate scrutiny test established in *Central Hudson* and applied in *R.J. Reynolds* and *Amestoy* would apply may depend on whether the compelled commercial speech solely aims at informing consumers (and is thus purely factual and uncontroversial) or whether the compelled commercial speech aims at altering consumer choice (and is, therefore, not purely factual and uncontroversial).[148]

4.3.3 Conclusion

The Supreme Court had not yet clarified the meaning of commercial speech, but corporate communications on food labels will likely be categorized as commercial speech. This opens the way for food businesses to challenge government regulations limiting corporate communications on food labels.

Cooked-up, *Corn Rebel*, *True-blue*, *My Goodness* and *Humble Honey* will likely be considered commercial speech. *Cooked-up* will probably not enjoy First Amendment protection as its claim that its canned macaroni and cheese dinner 'contains all the nutrients needed for a long and healthy life' is likely inherently misleading. The claims made by *Corn Rebel, True-blue, My Goodness*, and *Humble Honey*'s refusal to label its honey as being 'contaminated with genetically modified pollen' may be considered potentially misleading, provided the government can demonstrate a substantial interest, and may, therefore, require a disclaimer.

---

[146] *Discount Tobacco*, 674 F. 3d at 569.

[147] *R.J. Reynolds,* 696 F.3d at 1216.

[148] See for a more detailed analysis Keighley (2012), p. 539.

4.4 Comparing the Different Approaches to Free Speech Protection on Food Labels in the USA and the EU

I will compare the different approaches towards free speech protection on food labels in the EU and the US. First (1) I will compare the different approaches to whether corporate entities enjoy free speech protection. Second (2), I will compare whether food labels fall within the scope of free speech protection. Third (3), I will compare whether compelled expression enjoys free speech protection. Although food labels may enjoy free speech protection, this right may still be limited. I will, therefore, (4) assess the different approaches towards which limits can be set to free speech.

4.4.1 To Speak or Not to Speak: The Freedom of Companies to Express Themselves

Initially, in the USA the Amendments to the Constitution were seen as human rights, understood to apply to natural persons only. Over time, case law expanded the scope of the First, Fourth, Fifth, Sixth and Fourteenth Amendments to include corporate entities.[149] Within the EU certain fundamental rights, arguably including the freedom of expression, also extend to corporate entities.

4.4.2 Expression on the Food Label. Does the Food Label Have What It Takes?

In the EU and the USA expressions on food labels seem to enjoy free speech protection.

4.4.3 Can Humble Honey *Stay Humble? Whether Free Speech Includes the Right Not to Speak*

The extent to which free speech includes the right not to speak, such as *Humble Honey*'s refusal to label its honey as being 'contaminated with genetically modified pollen', differed per jurisdiction. In the USA compelled expression on food labels does enjoy free speech protection. However, purely factual and uncontroversial compelled expression on food labels aimed at informing consumers may be subject to a lighter review than compelled expression on food labels aiming at altering

[149] See Tucker (2001), p. 495 (and sources quoted there).

consumer choice. In the EU it is less clear whether free speech includes the right not to disclose information on food labels. Although the ECtHR found that journalists have a right not to speak, it is not self-evident that the same would apply to information on food labels as that type of information serves a fundamentally different purpose. Furthermore, the approach of the ECJ towards whether free speech includes a right not to speak is yet unclear.

4.4.4 Put the Lid On: How Free Speech on Food Labels Can Be Limited

To determine whether expressions can be limited, USA courts differentiate between types of expression: commercial speech has a limited First Amendment protection than political speech. The claims made by *Cooked-up*, *Corn Rebel*, *True-blue*, *Blueberrylicious, My Goodness,* and *Humble Honey*'s refusal to add a disclaimer will most likely be considered commercial speech. Which test will most likely be applied may depend on the type of commercial speech, mainly whether it concerns voluntary speech or compelled speech through disclaimers. *Central Hudson* test will likely be applied to *Cooked-up*, *Corn Rebel*, *True-blue*, *Blueberrylicious,* and *My Goodness.* This implies that commercial speech that is not misleading or unlawful may be limited when the government has a substantial interest to regulate the speech; when the limitation is necessary; and when the limitation is narrowly tailored to achieve the desired objective.[150] A lighter *Zauderer* test may be applied to disclosure requirements, such as that the government compels *Humble Honey's* to label its honey as being 'contaminated with genetically modified pollen'. The *Zauderer* test implies that the government may compel disclosure requirements associated with product marketing, so long as the disclosure is (1) purely factual and uncontroversial; (2) reasonably related to the State's interest in preventing deception of consumers; and (3) not unjustified or unduly burdensome.[151]

In the EU an overarching limitations clause applies to all types of expression. From the CFR, ECJ case-law, and ECtHR case-law it can be derived that limitations to free speech in the EU have to be provided for by law, must respect the essence of the fundamental right at issue, must serve a legitimate aim, and must be necessary.[152] The case-law of the ECtHR, which may be taken into account by the ECJ when assessing freedom of expression, differentiated commercial expression from other types of expression when it assessed the necessity of the limitation. Basically, the ECtHR granted the governmental authorities a wider margin of appreciation when assessing the necessity of a limitation to commercial expression compared to political expression. This margin may become narrower when the commercial

[150] Although this latest step seems to be eroding.

[151] *Zauderer v Office of Disciplinary Counsel of the Supreme Court of Ohio*, 471 U.S. 626 (1985).

[152] So-called principle of proportionality.

expression has 'political overtones'. It seems reasonable to expect that a similar approach as to the ECtHR will be used by the ECJ. Theoretically, it would also be conceivable that the ECJ, as a court of a union rooted in economic considerations, would value commercial expression higher than the ECtHR does. At present, however, there is no evidence pointing in this direction.

Food labelling regulations limiting free speech to protect public health seems to be generally regarded as legitimate to limit food labels in the analysed legal systems. Disclosure requirements seem to be always preferred in the USA and the EU over an outright prohibition as long as such a disclaimer has the same effectiveness as a prohibition.

4.5 Discussion: Food Businesses as Guardians of Sustainable Food information?

The degree to which free speech may be protected may differ depending on (1) whether the corporate nature of the speaker justifies free speech protection on food labels, or (2) whether it is the interest of the consumer, or (3) a public interest.[153] Practice in the EU and the USA shows that commercial statements enjoy limited protection compared to political statements. It is, therefore, important to consider (4) to what extent the nature of corporate communications on food labels may be considered political or commercial. Finally, I will consider (5) whether corporate free speech protection on food labels should include the right not to speak.

4.5.1 I Think, Therefore, I Am: The Corporate Nature of the Speaker as Justification for Corporate Free Speech

Most of the purposes and interests of free speech protection do not justify the protection of corporate speech on sustainability labels based on the corporate nature of the speaker.[154] Freedom of expression as a function of self-fulfilment and citizen participation in democracy seems to be a personal right that should only apply to human beings, as companies have no human dignity nor are capable of self-fulfilment. When corporate speech is political in nature, which may arguably be the case when it comes to sustainability labels, it can be argued that the tendency of

[153] See for a more general discussion on free speech protection Emerson (1963), p. 877 and Barendt (2011), n 3; For a more detailed discussion on corporate commercial free speech protection see Shiner (2003); Piety (2007), p. 2583.

[154] Ibid.

governments to suppress radical or subversive ideas might justify corporate free speech protection on food labels.

4.5.2 Give Me More: Consumer Interest to Receive Information as Justification for Corporate Free Speech

The corporate right to free speech could also be justified based on the consumer interest to receive the information.[155] If the food business is not allowed to give the information the consumer interest to receive the information is not met. It could be argued that the consumer interest to receive information demands that governments should not restrict corporate communication on food labels to protect the consumer right to make fundamental choices concerning their life, an important aspect of self-fulfilment.[156] This argument would only apply when consumers demand information, which the food business is not legally allowed to provide. Especially *Corn Rebel*'s claim that its sweet corn is being 'GM-free' enables consumers to only consume food products that fit within their lifestyle, i.e. foods that are not genetically modified. A similar argument would be more difficult to make for *True-blue*'s claim that Vitamin C has been added to its *Blueberrylicious* or *My Goodness*'s claim that 'the consumption of Bifidus eases the digestive system'. Furthermore, consumer interests to receive the information cannot justify a corporate right not to speak, so *Humble Honey*'s honey compelled disclosure that its honey is 'contaminated with genetically modified pollen' should be allowed.

4.5.3 The Truth, the Whole Truth, and Nothing but the Truth: Public Interest in the Free Flow of Information as Justification for Corporate Free Speech

The value of truth can be supported "by utilitarian considerations concerning progress and the development of society".[157] From this, it follows that the government should not regulate expression as this constitutes an interruption of the free flow of information. Interruption of the free flow of information has the potential that false information cannot be rebutted, which is harmful to society as a whole.

The free flow of false information, such as *Cooked-up*'s false claim that its canned macaroni and cheese dinner 'contains all the nutrients needed for a long and healthy life', is harmful to society: (1) false information increases the search costs for

---

[155] Barendt (2011), n 3, 25.

[156] Ibid, 401-2 referring to the Canadian Supreme Court decision *Ford v A-G of Quebec* [1988] 2 SCR 712, 767.

[157] Ibid, 7.

consumers to find good-quality goods; (2) the public might be incentivized to consume more canned macaroni and cheese dinners which is harmful to public health; and (3) food businesses will not be incentivized to innovate or improve their products to make truthful, non-misleading claims that appeal to consumers, because every competitor can make similar claims without it even being true.

This is different for claims that are in itself not false, such as *Corn Rebel*'s claim that its sweet corn is being 'GM-free', *True-blue*'s claim that Vitamin C has been added to its treats and *My Goodness*'s claim that 'the consumption of Bifidus eases the digestive system'. All seem to serve the public interest in the free flow of information i.e. lowering consumer search costs to make an optimal decision and promoting competition by stimulating the innovation and improvement of food products. In such a way both should also fall within the scope of free speech. Whether corporate information lowers consumer search costs to make an optimal decision can, however, be questioned as the amount of information consumers are exposed to is increased. This may result in information overload. Information overload actually increases search costs and could make it more difficult for consumers to process the information on the food label to make an informed choice of sufficient quality.[158] Consumer search costs to make an optimal decision might, therefore, be lowered with the limitation of the amount of information on food labels. If the government is not able to prove that consumers will be better enabled to make an optimal decision by withholding the information from the food label, the corporate information should fall within the scope of free speech protection.

4.5.4 Commercial or Political? The Nature of Corporate Communication on Food Labels

Some food businesses, such as *Corn Rebel*, might address topics through their food labels that are part of public debate, e.g. sustainability concerns such as that GMO/child labour/animal cruelty is bad, by informing the consumer that their food product is free from these qualities (e.g. *Corn Rebel*'s claim that its sweet corn is being 'GM-free'). It can be argued that the single statement that a product is free from qualities that might be considered bad by part of the public contributes as such to the public debate. There are much more obvious and more effective ways, however, to communicate political standpoints that certain qualities are bad than stating on food labels that the product does not contain that quality. This makes it difficult to argue that the main purpose of the corporate communication on the food label is political. Most food labels, including many sustainability labels, are, therefore, mainly commercial in nature as they are primarily targeted at affecting consumer purchasing decisions.

[158] Verbeke (2005), p. 347.

4.5.5 *Has* **Humble Honey** *a Free Speech Right Not to Speak?*

The right not to speak is "closely linked with freedom of belief and conscience and with underlying rights to human dignity, which would be seriously compromised by a legal requirement to enunciate opinions which are not in truth held by the individual."[159] As food businesses have no 'human dignity' it is hard to justify that food businesses have a free speech right not to disclose certain sustainability-related information. Furthermore, the consumer interest to receive information does not justify a corporate right not to disclose information on food labels. The public interest in the free flow of commercial information could, however, justify a corporate right not to disclose information when the food business can prove that the compelled information will increase consumer search cost to make an informed choice of sufficient quality. If the food business can prove this, the compelled claim should fall within the scope of free speech.

4.6 Conclusion

The communications made by *Cooked-up*, *Corn Rebel*, *True-Blue*, and *My Goodness* on their food labels will most likely enjoy, subject to limitations, free speech protection in the EU and the US. Such right can only be justified from the perspective of consumer or public interest. *Humble Honey* will most likely not enjoy a right to free speech in the EU, while this seems to be different in the US. Withholding information from consumers should, however, in general not fall within the scope of commercial speech protection because it generally does not serve the consumer or public interest in receiving information. This implies that *Humble Honey* has to speak up; at least as far as its free speech rights are concerned, except when *Humble Honey* can prove that the compelled information will increase consumer search costs to make an optimal decision.

Despite the right of free speech, the communications of *Cooked-up* can be banned in the USA and the EU because the government has a substantial interest in doing so. A complete ban of information may be necessary for utterly false claims, however, the claims made by *Corn Rebel*, *True-blue*, and *My Goodness* are not utterly false. Subject to substantial government interest, e.g. protecting public health, such claims can be prohibited, although it is most likely that a disclaimer will be required in the USA and the EU for these types of expressions as long as such a disclaimer will not increase consumer search costs to make an optimal decision.

In sum, food businesses should not be the guardians of sustainability information on food labels; their free speech rights on food labels should only be based on public and consumer interests, which could include sustainable consumption. Only when

[159] Barendt (2011), n 3, 94.

the information serves the public or consumer interests, free speech protection should step in.

References

Alexy R (2003) On balancing and subsumption. A structural comparison. Ratio Juris 16(433):440

Barendt E (2011) Freedom of expression. In: Rosenfeld M, Sajó A (eds) The Oxford handbook of comparative constitutional law. Oxford University Press, p 893

Biondi A (2012) Piet Eeckhout and Stefanie Ripley. In: EU law after Lisbon. Oxford University Press, p 163

Cannie H, Voorhoof D (2011) The abuse clause and freedom of expression in the European human rights convention: an added value for democracy and human rights protection. Neth Q Human Rights 29:54

Devenney J, Kenny M (2012) European consumer protection: theory and practice. Cambridge University Press, pp 349–350

Ehlers D (2007) European fundamental rights and freedoms. De Gruyter Recht, p 385

Emberland M (2006) The human rights of companies. Exploring the structure of ECHR protection. Oxford University Press, p 130

Emerson T (1963) Toward a general theory of the first amendment. Yale Law J 72:877

Gordley J (2012) The Functional Method. In: Monateri PG (ed) Methods of Comparative Law. Edward Elgar Publishing, p 113

Gottlieb S (1988) Compelling governmental interests: an essential but unanalyzed term in constitutional adjudication. Boston Univ Law Rev 58(917):948

Harris D, O'Boyle M, Warbick C (1995) Law of the European Convention on Human Rights. Butterworths, pp 378–379

Hirschl R (2005) The Question of Case Selection in Comparative Constitutional Law. Amerian Journal of Comparative Law 53:125

Keighley J (2012) Can you handle the truth? Compelled commercial speech and the first amendment. Univ Pa J Const Law 15:539

Piety T (2007) Against freedom of commercial expression. Cardozo Law Rev 29:2583

Purnhagen K (2014) The Virtue of Cassis de Dijon 25 years later—it is not dead, it just Smells Funny. In: Purnhagen K, Rott P (eds) Varieties of European economic law and regulation. Springer, pp 329–332

Rengeling H, Szczekalla P (2004) Grundrechte in der Europäischen Union: Charta der Grundrechte und allgemeine Rechtsgrundsätze. Heymanns, pp 344–390

Schorkopf F (2007) Human dignity, fundamental rights of personality and communication. In: Ehlers D (ed) European fundamental rights and freedoms. De Gruyter Recht, p 425

Shiner R (2003) Freedom of commercial expression. Oxford University Press

Trstenjak V, Beysen E (2013) The growing overlap of fundamental freedoms and fundamental rights in the case-law of the ECJ. Eur Law Rev 38:293

Tucker A (2001) Flawed assumptions: a corporate law analysis of free speech and corporative personhood in citizens United. Case West Reserve Law Rev 61:495

Verbeke W (2005) Agriculture and the food industry in the information age. Eur Rev Agric Econ 32:347

Weatherill S (2011) The limits of legislative harmonisation ten years after Tobacco advertising: how the Court's case law has become a "Drafting Guide". German Law J 12:827

Weiβ W (2012) EU human rights protection after Lisbon. In: Trybus M, Rubini L (eds) The treaty of Lisbon and the future of European law policy. Edward Elgar Publishing, p 224

Chapter 5
Sustainability Labels in the Shadow of EU Competition Law

5.1 Introduction

In recent years, agreements between undertakings to stop the sale of unsustainable products or production processes have been discouraged by national competition authorities (NCAs) in the EU. NCAs argued in these cases that the objective benefits (meaning quantifiable benefits and not the subjective opinion of the undertakings)[1] to the consumers of the product were lower than the costs of the agreement.[2] The NCAs seemingly based their decision on their interpretation of the European Commission (hereinafter: Commission) guidelines issued in 2004 on the application of Article 101(3) of the Treaty on the Functioning of the European Union (TFEU)

This book chapter is based on Eva van der Zee, 'Quantifying Benefits of Sustainability Agreements under Article 101 TFEU' (2020) 43(2) World Competition Law and Economic Review.

[1] The term "objective" has different connotations depending on who uses the word. The Commission sees "objective" benefits as meaning empirically quantifiable, and not the subjective opinion of undertakings. "Objective" benefits are used in a different way by economists, psychologists, and moral philosophers as will be apparent in this paper. Economists, psychologists, and moral philosophers see "objective" benefits as resources or capabilities that are determined to be important by a society as a whole (often laid down as fundamental rights, or as material resources). These "objective" benefits are, therefore, not necessary quantifiable—although great progress have emerged in this regard. "Subjective" benefits are, however, quantifiable, as they are empirical data derived from asking or studying individual's subjective perception of happiness or subjective preferences.

[2] Autoriteit Consument en Markt, 'Analyse ACM van Duurzaamheidsafspraken 'De Kip van Morgen" (2015) <https://www.acm.nl/nl/download/publicatie/?id=13758>; See also Dolmans (2020), p. 17 for a discussion on ACM's approach in the Kip van Morgen case; Conseil de la Concurrence. Avis relatif à l'examen, au regard des règles de concurrence, des modalités de fonctionnement de la filière du commerce équitable en France. N° 06-A- 07 du 22 mars 2006, para 97.

© The Author(s), under exclusive license to Springer Nature Switzerland AG 2022
E. van der Zee, *Sustainability Labels in the Shadow of the Law*, Studies in European Economic Law and Regulation 22, https://doi.org/10.1007/978-3-030-95802-2_5

(hereinafter Commission guidelines).[3] The Commission guidelines are often interpreted as that only benefits that can be subsumed within the notion of economic efficiency can be taken into account, meaning that non-economic justifications cannot (with the term "economic" seemingly only referring to monetary values).[4] Such an interpretation of the Commission guidelines seems to be logical at first sight. For example, the Commission observed in other documents that the purpose of Article 101(3) TFEU is to provide a legal framework for the economic assessment of restrictive practices.[5] In addition, the Commission states that the purpose of the guidelines is to develop a method for the application of Article 101(3) TFEU based on the economic approach introduced and developed in the guidelines on vertical restraints, horizontal co-operation agreements, and technology transfer agreements.[6] With this more "economic approach" the Commission refers to its interpretation of the aim of EU competition rules: enhancing consumer welfare and ensuring an efficient allocation of resources,[7] which are two concepts derived from economic theory.

However, the Commission is far from clear on what it means with the term "efficiency"[8] nor what it means with the concept of "consumer welfare".[9] This is problematic because the definition of these terms determines whether and to what extent considerations that cannot be (easily) monetized (such as some aspects of sustainable development) can be taken into account under the analysis. It has often been argued that only benefits with a clear monetary value can be taken into account under Article 101(3) TFEU.[10] However, the Commission never made it explicit in its guidelines that only benefits with a clear monetary value could be taken into account. In fact, the guidelines do not clarify how to quantify benefits under the competition rules at all, or that they should be quantified solely in monetary terms. Due to this

---

[3] European Commission, "Notice: Guidelines on the application of Article 81(3) of the Treaty" [2004] OJ C 101/97, 2004, at 84.

[4] See for a discussion on the matter Gerbrandy (2017), p. 539; Whish and Bailey (2015), p. 140; Monti and Mulder (2017); Article 81(3) Guidelines, n 3; Odudu (2006); Odudu (2010), p. 599; Kingston (2012); Townley (2009); Casey (2009), p. 362; Gerbrandy (2013); Nowag (2016); Monti (2020).

[5] European Commission, 'White Paper on the Modernisation of the Rules Implementing Articles 85 and 86 of the EC Treaty', (1999). The approach has been confirmed in the European Commission, 'Guidelines on the applicability of Article 101 of the Treaty on the Functioning of the European Union to horizontal co-operation agreements' (2011) OJ C11/ and the European Commission, 'Guidelines on Vertical Restraints', (2010) 2010/C 130/01.

[6] Ibid, at para 5.

[7] Ibid, at para 33.

[8] There are various sources of efficiency gains: Pareto, static (allocative or productive), and dynamic. These sources of efficiency gains may potentially conflict. See for more information on sources of efficiency gains Van den Bergh (2017).

[9] Daskalova (2015), pp. 131–160.

[10] See for a discussion on the matter Whish and Bailey (2015), n 4; Monti and Mulder (2017), n. 4; Article 81(3) Guidelines, n 3; Odudu (2006), n. 4; Odudu (2010), n. 4; Kingston (2012), n. 4; Townley (2009), n. 4; Casey (2009), n. 4; Gerbrandy (2013), n. 4; Nowag (2016), n. 4.

lack of clarity of the guidelines and due to the resulting interpretation by NCAs,[11] undertakings have been hesitant to set up new sustainability agreements out of fear of competition rules.[12]

The main aim of this chapter is to examine how the current Commission guidelines could be improved to allow undertakings to assess their agreements under Article 101(3) TFEU in a way that is quantifiable but that goes beyond the economic approach focusing solely on monetary well-being (hereinafter: welfare economic approach). Defenders of the welfare economic approach have long argued that, despite its flaws, a welfare economic approach is the best available means for determining the efficiencies and benefits of anti-competitive agreements.[13] That may no longer be the case. This paper will propose an alternative method for analysing the efficiencies and benefits of an anti-competitive sustainability agreement: human well-being (HWB).

To achieve this aim, I will provide a brief overview of Article 101 TFEU and the Commission guidelines, pointing out where further guidance is needed (Sect. 5.2). Section 5.3 will show that based on communications by the Commission, as well as based on case law by the Court of Justice of the European Union (CJEU), the term "benefits" under Article 101(3) TFEU should be interpreted as increasing a wide notion of well-being (i.e. not solely related to monetary well-being), also known as HWB. Section 5.4, will discuss different methods to estimate HWB (i.e. economic cost-benefit analysis, well-being analysis, resourcism, and the capabilities approach), arguing that a combination of these methods would provide the most accurate estimation of HWB. Section 5.5 will provide further information on how to incorporate HWB in the analysis under Article 101 TFEU.

5.2 Article 101 TFEU and Its Possible Exceptions

Article 101(1) TFEU prohibits agreements between undertakings that may affect trade between Member States and which have as their objective or effect the prevention, restriction or distortion of competition within the internal market. For an agreement to be caught by Article 101(1) TFEU depends firstly on whether the agreement is between undertakings. According to the CJEU 'the concept of an undertaking encompasses every entity engaged in an economic activity, regardless

---

[11] Autoriteit Consument en Markt (2014), n 2; Conseil de la Concurrence. Avis relatif à l'examen, au regard des règles de concurrence, des modalités de fonctionnement de la filière du commerce équitable en France. N° 06-A- 07 du 22 mars 2006, para 97.

[12] The author interviewed several companies and NGOs that indicated their hesitance to take steps towards more sustainable production out of fear of competition rules. More details can be obtained from the author.

[13] This has especially been argued for economic cost-benefit analysis. See e.g. Adler and Posner (2006).

of the legal status of the entity and the way in which it is financed'.[14] The main controversial criterion when determining whether the actors involved in the agreement are undertakings, is that an undertaking needs to carry out the activity to make profits.[15] This criterion would in any case exclude sustainability agreements between undertakings that carry out activities as a task in the public interest typical of public authority,[16] or when they fulfil a social function that is predominantly based on solidarity.[17]

Promoting sustainable development is often not the sole objective of undertakings as ultimately they aim to make profits by the agreement, e.g. by preventing that consumers would switch to the competitor for cheaper, less sustainable, products.[18] Therefore, only agreements between undertakings that ultimately aim to make profits by the agreement will be considered in this chapter. As such, the solidarity argument or the useful effect doctrine is assumed not to apply in these cases.[19]

Undertakings need to assess themselves whether their agreement infringes Article 101(1) TFEU and, if so, whether the exemption conditions of Article 101(3) TFEU are met. Four conditions set forth in Article 101(3) TFEU must be met to justify agreements caught by Article 101(1) TFEU: the agreement should (1) contribute to the improvement of the production or distribution of goods, or to promote technical or economic progress, (2) while allowing consumers a fair share of the resulting benefits and (3) which does not impose non-indispensable restrictions on the undertakings concerned and (4) which does not afford these undertakings the possibility of eliminating competition. A fifth condition, added by the CJEU, requires that the benefits that the agreement yields must be greater than its costs.[20]

---

[14] Case 41/90 *Höfner and Elser v Macrotron*, ECLI:EU:C:1991:161, at 21; Case 280/06 *ETI and Others*, ECLI:EU:C:2007:775, at 38; Case 350/07 *Kattner Stahlbau*, ECLI:EU:C:2009:127, at 34; see also Odudu (2006), n 4, 26 and 214; Whish and Bailey (2015), n 4, 88–89.

[15] Case 343/95 *Diego Cali & Figli*, ECLI:EU:C:1997:160.

[16] Case 107/84 *Commission v Germany*, ECLI:EU:C:1985:332, at 14-15; Case 364/92 *SAT Fluggesellschaft v Eurocontrol*, ECLI:EU:C:1994:7, at 27-30; Case 67/96, Joined Cases 115/97 to 117/97, and Case 219/97 *Albany, Brentjens, Maatschappij*, ECLI:EU:C:1999:28; Case 159/91 and 160/91 *Poucet and Pistre*, ECLI:EU:C:1992:358 at 12.

[17] Gerbrandy (2017), n 4.

[18] It should be noted that supermarkets do not always embark on horizontal co-operation agreements to stop the sale of unsustainable alternatives. For example, Dutch supermarkets individually agreed to stop the sale of battery eggs. "Supermarkten weren eieren uit legbatterij", De Volkskrant, June 6, 2003, p. 15. Arguably, the reason why in this case a horizontal co-operation agreement was not deemed to be necessary was because consumers do not select a supermarket based on the price of eggs, but they do select a supermarket based on the price of meat. As such, supermarkets may be more affected when increasing the price of meat. See e.g. DeGraba (2006), pp. 613–628.

[19] It goes beyond the aim of this paper to discuss the solidarity argument or the useful effect doctrine. For further information on how the solidarity argument and the useful effect doctrine could apply to sustainability agreements under Article 101 TFEU, see Gerbrandy (2017), n 17.

[20] Monti and Mulder (2017), n 4, 647 referring to Case 56/64 *Etablissements Consten Sàrl v Commission*, ECLI:EU:C:1966:41 at 348–349.

The Commission issued guidelines in 2004 on the conditions for exception contained in Article 101(3) TFEU, to guide undertakings in their assessment.[21] The Commission guidelines to Article 101(3) TFEU do not specifically clarify which method to use to quantify the claimed efficiencies and the resulting benefits to the consumers of an agreement. Some guidance could be found in the 2001 Horizontal Cooperation Guidelines in which the Commission held that:

> Environmental agreements caught by Article 81(1) [now Article 101(1)] may attain economic benefits which, either at individual or aggregate consumer level, outweigh their negative effects on competition. To fulfil this condition, there must be net benefits in terms of reduced environmental pressure resulting from the agreement, as compared to a baseline where no action is taken. In other words, the expected economic benefits must outweigh the costs.
>
> Such costs include the effects of lessened competition along with compliance costs for economic operators and/or effects on third parties. The benefits might be assessed in two stages. Where consumers individually have a positive rate of return from the agreement under reasonable payback periods, there is no need for the aggregate environmental benefits to be objectively established. Otherwise, a cost-benefit analysis may be necessary to assess whether net benefits for consumers in general are likely under reasonable assumptions.[22]

The 2001 Horizontal Guidelines have been replaced by the 2010 Horizontal Guidelines that do no longer include a similar reference to environmental agreements and cost-benefit analysis (CBA).[23] The Commission has explained this is not to be seen as a change of its policy.[24] As such, the 2001 Horizontal Guidelines provide some guidance, namely that a CBA is warranted, although it does not clarify what type of CBA should be conducted and whether it should only be assessed in monetary terms.

Some scholars argue that sustainability benefits that go beyond monetary gains have no place in the legal analysis under Article 101(3) TFEU.[25] These scholars point to the Wouters doctrine to nonetheless allow sustainability benefits that go beyond monetary gains in the assessment under Article 101 TFEU.[26] In Wouters the CJEU drew upon its jurisprudence on internal market law, arguing that the decision of an association of undertakings (the Dutch Bar Association) which restricted

---

[21] Article 81(3) Guidelines, n 3.

[22] Commission Guidelines on the Applicability of Art. 81 of the EC Treaty to Horizontal Co-Operation Agreements [2001] OJ C 3 (6 Jan. 2001), para 193-4.

[23] European Commission, "Guidelines on the applicability of Article 101 of the Treaty on the Functioning of the European Union to horizontal co-operation agreements" [2011] OJ C11.

[24] Commission, Competition: Commission Adopts Revised Competition Rules on Horizontal Cooperation Agreements – Frequently Asked Questions 4 (2010), https://europa.eu/rapid/press-release_MEMO-10-676_en.htm?locale=en.

[25] Case 309/99 *Wouters*, ECLI:EU:C:2002:98.

[26] Whish and Bailey (2015), n 14, 140; Monti and Mulder (2017), n 4; Article 81(3) Guidelines, n 4; European Commission, "Guidelines on the applicability of Article 101 of the Treaty on the Functioning of the European Union to horizontal co-operation agreements" [2011] OJ C11/1; Odudu (2006), n 14; Monti and Mulder (2017), n 4; Kingston (2012), n 4; Townley (2009), n 4; Casey (2009), n 4; Gerbrandy (2013), n 4; Nowag (2016), n 4.

competition (banning multi-disciplinary practices[27]) was necessary to achieve a public interest objective (the proper practice of the legal profession). In Wouters, the CJEU held that the restriction of competition was suitable (the restriction could reasonably be considered necessary to achieve its public interest objective[28]), necessary (the Netherlands was entitled to consider that the objectives pursued by the restriction cannot be attained in less restrictive means[29]), and did not have a disproportionate effect on other interests (the effects of the restriction does not go beyond what is necessary to ensure the public interest objective[30]). As such, in effect, the CJEU conducted a proportionality test.[31]

It is often mentioned that under the Wouters doctrine, contrary to Article 101(3) TFEU, the agreement does not have to benefit the individual consumer.[32] However, in the pursuit of sustainable development, individual consumer and societal interests are often aligned.[33] For example, an individual consumer may immediately benefit from cheaper product prices, but these cheaper product prices may undermine broader social welfare (relocation of production to less expensive labour markets, increasing environmental pollution due to longer distance these goods must travel), and ultimately circles back to the detriment of the consumer (unemployment; environmental pollution).[34] In other words, an agreement that may be exempted following, for example, the Wouters doctrine when it pursues societal goals, would often also come to benefit the consumer in the long term, especially as, as held by the Commission in CECED, these benefits to the consumer may be rather small.[35]

Second, some argue that the Wouters doctrine should be preferred over an analysis under Article 101(3) TFEU when assessing sustainability agreements since it does not require quantification of benefits.[36] However, this argument leaves aside the general trend in the internal market case law of the CJEU to leave open the possibility for Member States to use statistical evidence or any other suitable type of

---

[27] Case 309/99 *Wouters*, ECLI:EU:C:2002:98, at 100.

[28] Ibid, at 107.

[29] Ibid, at 108.

[30] Ibid, at 109.

[31] For a proportionality test to be satisfied the restriction must (1) be a useful, suitable, or effective means of achieving a legitimate aim; (2) not be more restrictive than necessary to achieve that legitimate objective; (3) not have an excessive or disproportionate effect on the other interests. De Burcá (1993), p. 105.

[32] Monti and Mulder (2017), n 4; Case 439/09 *Pierre Fabre*, ECLI:EU:C:2011:113 at 35; Jansen and Kloosterhuis (2016), p. 335.

[33] Glen et al. (2012), p. 257.

[34] Hardin (1968), p. 1243; Platt (1973), p. 641; Schelling (1984), p. 1; Beck (1992); Beck (2014).

[35] *CECED*, OJ 2000 L 187/47, at 53.

[36] See e.g. Mortelmans (2001), p. 613ibid; Monti (2002), p. 1057; Townley (2009), n 4; Casey (2009), n 4; Kingston (2012), n 4; Jan Ankersmit (2017), p. 194; Nowag (2016), n 4; Monti and Mulder (2017), n 4.

evidence when conducting the proportionality test.[37] As the Wouters doctrine is based on the CJEU's jurisprudence on internal market law,[38] it possibly warrants a similar approach. Further clarity on how to select and weigh statistical evidence under the Wouters doctrine would, therefore, also be useful for NCAs, governments, and courts to quantify the benefits of an agreement.

To provide undertakings and governments with some further guidance, I propose that undertakings conduct a CBA based on human well-being to quantify the claimed efficiencies and the resulting benefits to the consumers of an agreement. HWB is a complex concept as it embraces many different ideas and perspectives and many discussions on HWB take place within philosophy, psychology, and economics. There are generally two approaches underlying HWB: subjective well-being or utility (i.e. happiness, preferences) and objective well-being (i.e. material resources, capabilities). For a long time, it was only possible to measure HWB in terms of preferences or material resources. So, subjective well-being was often measured by asking individuals about their preferences in monetary terms (stated preferences) or by distilling these preferences through their revealed behaviour (revealed preferences). Objective well-being was measured by assessing solely material resources that come with a price tag (such as income and resources). However, scholars have developed techniques to measure happiness and capabilities as well in comparable units that go beyond mere preferences or material resources.[39] As such, it is now possible to measure HWB more precisely and balance the different aspects of well-being against each other in a CBA.

HWB gives tremendous opportunities in law, where balancing conflicting norms and case-specific circumstances are often pivotal to reaching acceptable outcomes.[40] In addition, HWB would be very helpful in EU competition law as it provides more detailed guidance to undertakings in their self-assessment under Article 101 TFEU. Moreover, as will be shown in the next paragraph, using HWB as a tool to measure benefits under Article 101(3) TFEU is more in line with the welfare standards proposed by the Commission and the CJEU.

5.3 The Welfare Standard Under EU Competition Law

According to the Commission, the aim of EU competition law is to enhance consumer welfare and ensuring an efficient allocation of resources.[41] The consumer welfare standard could either be narrowly defined (where only price increases are

---

[37] Case 333/14 *Scotch Whiskey*, ECLI:EU:C:2015:845, at 64-5 in conjunction with Case 148/15 *Deutsche Parkinson Vereinigung*, ECLI:EUC:2016:776, at 35-6.

[38] Monti and Mulder (2017), n 4, 645.

[39] See e.g. Anand and van Hees (2006), pp. 268–284.

[40] Gerbrandy (2017), n 4, at 553, referring to Alexy (2003), p. 131; Klatt and Meister (2012).

[41] Ibid, at para 33.

considered) or more broadly (where efficiencies as well as non-price considerations, e.g. those related to product quality, choice, and innovation, can be taken into account to offset an ex-post price increase).[42] While consumer welfare is often understood as a welfare economic notion only taking into account price considerations,[43] this has never been clarified as such by the Commission.[44] There are, in fact, two reasons to assume that the Commission applies a broad consumer welfare standard.

First, the Commission explicitly recognises the role that dynamic efficiencies can play in assessing anti-competitive agreements.[45] Dynamic efficiency involves improving allocative efficiency (ensuring resources are allocated between alternative uses in a way that maximises wellbeing) and productive efficiency (output is produced at its lowest possible average costs) over time.[46] This can be done, e.g., by developing new or better products and finding better ways of producing goods and services. Following this rationale, anti-competitive agreements can achieve dynamic efficiency when they yield lower-cost or higher-quality outputs to the benefit of total welfare. These lower-costs or higher-quality outputs can, then, benefit the consumer. This means that anti-competitive agreements that may raise prices but improve quality, may still be justified under Article 101(3) TFEU according to the Commission, when increased quality benefits the consumer. This is important for sustainability agreements, as they often increase the quality of the output (e.g. less pesticide residue, increased animal welfare, better labour conditions).

Second, the Commission argues in its Better Regulation Guidelines (2015) that not every situation calls for a narrow assessment to measure consumer welfare, i.e. measured by the amount a consumer is willing to pay (WTP) for a good minus the amount a consumer actually pays for it. The Commission argues that sometimes it is difficult to quantify impacts and sometimes further information is needed on how consumer welfare may change over time.[47] Consumer WTP is particularly difficult to measure for goods that do not have a clear price tag, such as sustainable development. In cases where it is difficult to quantify impacts, the Commission proposes to survey the personal experience of consumers through surveys on financial and non-financial detriment such as time losses and psychological detriment.[48] Hereby, the Commission acknowledges the weakness of resorting solely to price considerations to assess consumer welfare by emphasizing the importance of using behavioural insights in the assessment of consumer welfare, acknowledging that consumers' limited, potentially biased, and socially influenced decision-making

[42] See on the consumer welfare standard more generally Browning and Zupan (2009).

[43] See for example: Townley (2011), p. 441; Kingston (2012), n 4.

[44] See e.g. Daskalova (2015), n 9, 131–160.

[45] Article 81(3) Guidelines, n 3.

[46] Browning and Zupan (2009), n 42.

[47] European Commission, "Better Regulation Guidelines", Strasbourg, 19.5.2015, toolbox #32, at 253-4.

[48] Ibid.

affects how they make choices in the marketplace. In addition, by stating that also psychological detriment should be surveyed,[49] the Commission hints that in some cases a well-being analysis (WBA), where the happiness of consumers is assessed, is appropriate than solely resorting to a narrow assessment of consumer welfare.

Although the Better Regulation Guidelines (2015) are developed to guide the Commission in preparing and evaluating public policy, it may still be useful as a guide for our current purpose as it provides the current understanding by the Commission on how consumer welfare should be measured.

The CJEU also seems to apply a broad welfare standard. In its 2011 judgment in Telia Sonera, the CJEU held that the function of competition rules is 'to prevent competition from being distorted to the detriment of the public interest, individual undertakings and consumers, thereby ensuring the well-being of the European Union.'[50] The CJEU thereby referred to earlier case-law in Roquette Frères from 2002 in which it stipulated that the function of competition rules is to ensure the 'economic well-being in the Community'.[51] The judgment of the CJEU in Telia Sonera should be read as not referring solely to 'standard of living', but also including quality of life. Roquette Frères concerned a referral for a preliminary ruling concerning the EEC Treaty. At the time of the EEC Treaty, the aim of the Community was limited to promoting a standard of living.[52] Telia Sonera concerned a preliminary ruling concerning the Treaty of Lisbon, in which the overall aim of the European Union was expanded to promote not only the standard of living but also quality of life.[53] The term 'standard of living' (Article 2 ECSC Treaty, later Article 2 EEC Treaty) is solely a measure of monetary well-being or "the degree of wealth and material comfort available to a person or a community".[54] 'Quality of life' is "the standard of health, comfort, and happiness experienced by an individual or group".[55] Quality of life is defined by Eurostat, the directorate-general of the Commission responsible for providing statistical information to the institutions of the European Union, as a broad concept encompassing objective well-being (such as a standard of living, income, health) and the subjective perception one has of them.[56]

[49] Ibid.

[50] Case 52/09 *TeliaSonera Sverige*, ECLI:EU:C:2011:83, at 22.

[51] Case 94/00 *Roquette Frères*, ECLI:EU:C:2002:603, at 42.

[52] Art. 2 EEC Treaty.

[53] Art. 3(1) TEU emphasises that the overall aim of the European Union is "to promote (...) the well-being of its peoples." It should be noted that the term 'well-being' inArt.3(1) TEU should be understood in a broader sense than the economic and formerly used term 'standard of living' (Art. 2 ECSC Treaty, later Art. 2 EEC Treaty) for two reasons. First, Recital 9 of the Preamble of the TEU clearly states that the EU is determined to promote economic and social progress for their peoples. Second, Art. 3 TEU is based on the former Art. 2 EC which states that the aim is not only to raise the 'standard of living' but also to raise the 'quality of life'. As such, the term 'well-being' should be understood in a broader sense including quality of life.

[54] Soanes and Stevenson (2005).

[55] Ibid.

[56] Eurostat, Quality of Life Indicators—Measuring quality of life, https://ec.europa.eu/eurostat/statistics-explained/index.php?title=Quality_of_life_indicators_-_measuring_quality_of_life.

As the Commission and the CJEU favour a broad welfare standard that goes beyond monetary well-being, HWB could provide important insights to quantify the benefits of anti-competitive agreements under Article 101 TFEU in a way that goes beyond monetary well-being. The methods underlying such analysis will be explored in the next section.

5.4 Methods to Estimate Human Well-Being

> What we measure affects what we do; and if our measurements are flawed, decisions may be distorted.[57]

To estimate HWB, researchers have combined welfare economics, social psychology, and moral philosophy.[58] This approach adds to the approach by welfare economics (where benefits are estimated in monetary terms) by adding objective and subjective factors of well-being that do not have a clear monetary value. The remainder of this chapter will address the different aspects of measuring HWB, its strengths and weaknesses, and how statistical data derived from this approach could be helpful in the analysis under Article 101 TFEU.

5.4.1 Subjective Well-Being

Subjective concepts and measures of HWB view well-being in terms of utility, i.e. happiness or the satisfaction of preferences. It is based on classic utilitarianism, of which the roots are often traced back to moral philosopher and legal reformer Jeremy Bentham. Bentham observes that:

> Nature has placed mankind under the governance of two sovereign masters, pain and pleasure. It is for them alone to point out what we ought to do. . .[59]

As such, well-being is measured according to the duration and intensity of pleasure and pain. There have been many attempts to measure utility in economics and psychology. The two most well-known methods are economic CBA and subjective well-being analysis (WBA).

[57] Stiglitz, Sen and Fitoussi, *Report by the Commission on the Measurement of Economic Performance and Social Progress*.

[58] Ibid.

[59] Originally published in 1789, Bentham (2007).

5.4.1.1 Economic Cost-Benefit Analysis

Economic CBA surveys the utility individuals derive from markets or policy measures by measuring whether individual preferences are satisfied. They derive information about individual preferences from stated preferences and revealed preferences. Stated preferences, also known as contingent valuation, is a questionnaire-based technique that asks individuals what they would be willing to pay (WTP) or what they would be willing to accept (WTA) to undergo or to avoid a specific change of the status quo.[60] Revealed preferences are available when individuals have had the opportunity to choose between a change of the status quo and an amount of money to retain the status quo, such that it can be simply observed which option they chose. Stated or revealed preferences is said to reveal whether the individual prefers, for example, having more money or reduced environmental pollution. Both stated and revealed preferences are converted into a uniform metric, monetary value.[61]

An important strength of economic CBA is the possible gain in transparency by forcing decisions to be based on rigorous analysis made available to the public.[62] Another strength is that businesses likely emphasise the costs of the agreement in monetary terms. When sustainability benefits, then, cannot be compared with these costs, as there is no similar (monetary) unit of comparison, this will lead to lower protection of those values.[63]

Economic CBA is, like any methodology, subject to limitations.[64] An important limitation of economic CBA is that all costs and benefits need to be converted into monetary value. This may be difficult for benefits that do not have a clear market price,[65] such as future and societal benefits of sustainability agreements. The outcome of a CBA in which future and societal benefits are considered may therefore differ depending on the value given to future benefits over present benefits, and the extent to which human beings are considered to be subject to biases. Awareness of these limitations is essential to accurately interpret the outcome of an economic CBA and its value to inform legal analyses under EU competition law.

To estimate future benefits in monetary terms, welfare economists convert future costs and benefits to present value to make them comparable with each other.[66] To convert future costs and benefits to present value, welfare economists use a discount rate.[67] This is realised through the mechanism known as discounting. Discounting

[60] Browning and Zupan (2009), n 42.

[61] Bronsteen (2013), p. 1603.

[62] Wiener (2006), pp. 447–518.

[63] Buchanan (1987), pp. 718–21.

[64] Sunstein (2018); Wiener (2006), n 62, 447.

[65] Ibid.

[66] See e.g. Feldstein (1964), pp. 294, 360–379; Keating and Keating (2017) chapter 3; Nas (2016), Chapter 12.

[67] Olson and Bailey (1981), p. 1.

means that lower weight is put on future costs and benefits than on present costs and benefits.[68] The choice of a discount rate remains a highly political matter as it reflects society's preferences between present and future consumption. For example, the higher the discount rate the more the costs of environmental degradation are shifted to future generations. Statistical data based on the use of a discount rate should, therefore, be interpreted with awareness of which discount rate is applied and for what underlying reasons.[69]

In addition, economic CBA is based on the idea that how much money a person is willing to pay or accept for a commodity shows how much the commodity increases their well-being. But this may not always be true. Studies have demonstrated that individuals are not very good at predicting how their choices will affect their well-being in the future.[70] Insights from subjective WBA may, therefore, further add to our understanding of the utility an individual derived from a change of the status quo.[71]

5.4.1.2 Well-Being Analysis

Social psychologists use subjective WBA to survey the utility individuals derive from markets or policy measures by measuring the happiness of individuals. They derive information from hedonic psychology data to measure the effects of a change of the status quo on people's happiness. The data is derived from individuals' personal assessments of how they feel at a particular moment in time. Psychologists and economists have developed surveying and statistical methods that enable the collection and analysis of hedonic data on a large scale.[72] The effects of different factors on people's life satisfaction or happiness are not converted in monetary values but into subjective, hedonic, cardinal, and interpersonally comparable units.[73]

WBA encompasses different aspects, such as cognitive evaluations of one's life, positive emotions such as joy and pride, and negative emotions such as pain and worry.[74] When a sustainability agreement results in less pollution but higher prices, WBA can compare people's happiness with less pollution with their happiness with decreased buying power.

---

[68] Keating and Keating, n 66.

[69] For further reading: Padilla (2002), pp. 69–83; Gollier (2002), pp. 2, 463–473; Norgaard and Howarth (1991), pp. 88–101.

[70] Gilbert and Wilson (2007), p. 1351; Wilson and Gilbert (2005), p. 131.

[71] Sunstein (2018).

[72] Kahneman et al. (1999).

[73] Bronsteen et al. (2013), n. 61, 1618.

[74] Stiglitz et al. (2009), p. 33.

The main strength of WBA is that WBA is less subject to cognitive biases that may occur in economic CBA, such as people's difficulty to predict the impact of their choices or life events on the future.[75]

An important pitfall of WBA is that it uses evaluative judgments of a person's positive and negative feelings in real-time (i.e. hedonic experiences). Therefore, WBA excludes preferences of future consumers or, similarly as in economic CBA, discounts these preferences. Another pitfall of WBA is that it fails to consider what is necessary to live a human life as people may adapt to their life circumstances.[76] Studies show, for example, that individuals who lose limbs often adapt to their new condition, recovering most of their lost happiness with a few years.[77] As such, a change of the status quo that may be considered problematic from a moral point of view (e.g. people will use limbs due to environmental pollution), could be considered less problematic from the perspective of WBA.

5.4.2 *Objective Well-Being*

Although subjective well-being is an important factor of well-being, it should not be treated as 'general-purpose guides to all aspects of well-being'.[78] The objective well-being approach assumes that people have basic needs and rights, ranging from adequate income to being able to play. Using this approach, HWB can be assessed through the analysis of objective indicators (meaning not the subjective preferences or feelings of individuals subject to the change of the status quo). There are two theories underlying objective well-being: resourcism and the capabilities approach.

5.4.2.1 Resourcist Approach

The resourcist approach defines welfare by looking at each person's possession of specific resources.[79] An example of a resourcist approach is gross domestic product (GDP) per capita where the progress of society's are defined by the average resource bundle per person.[80] The resourcist approach has been widely criticised as it fails to take into account resources that do not have a clear price tag (e.g. natural resources) and it fails to take into account that given differential personal, social, and

---

75 Adler and Posner (2008), Chap. 1, n 32. Gilbert and Wilson (2007), n 70.

76 See e.g. Luhmann et al. (2012), p. 592.

77 For a summary of the initial research on hedonic adaptation, see Frederick and Loewenstein (1999), p. 302.

78 Sen (2009), p. 286.

79 See on resourcism more generally: Rawls (1982), pp. 86–87; Pogge (2005), p. 1.

80 Claassen and Gerbrandy (2016), p. 1 at 5.

environmental conversion factors, people may require different resources to achieve similar outcomes of well-being.[81]

Welfare economists and psychologists have attempted to circumvent the latter problem, by measuring well-being more directly by estimating the utility individuals derive from specific resources. As described above, subjective well-being methods, although valuable, are limited in the sense that individuals are not very good at predicting how their choices will affect their well-being in the future,[82] and because it can miss the urgency of specific consequences when certain resources get lost as people may adapt to their life-circumstances.[83]

5.4.2.2 Capabilities Approach

Amartya Sen and Martha Nussbaum have developed the capability approach to propose an alternative view of welfare that goes beyond utilitarian and resourcist theories.[84] A capability is an ability or opportunity to function in a certain way.[85] The capability approach compares a person's well-being to that of others based on those capabilities that are considered important in human life according to a predetermined list, instead of on preferences or happiness.[86]

The capability approach may provide additional insights to conduct a more thorough analysis of the impact of an agreement on objective well-being. The strength of including the capability approach is that it includes important values for well-being besides material resources, and what people desire or that what they consider to make them happy.

To use the capability approach, we need projections or prior responses to normative questions as to what capabilities are considered important in human life. The selection of relevant capabilities is a value judgment of which capabilities we consider important. Different ideas as to how such capabilities should be determined include a pre-existing list of central capabilities[87] and leaving the selection of capabilities to democratic processes.[88] Often, statisticians use the list of central capabilities provided by Nussbaum to develop statistical indicators.[89] These central

[81] See e.g. Nussbaum (2000); Sen (2009), n 78; Anderson (2010).

[82] Gilbert and Wilson (2007); Wilson and Gilbert (2005), n 70.

[83] See e.g. Luhmann et al. (2012), n 76.

[84] Sen (1999); Sen (1987), p. 17; Nussbaum (1988), p. 145; Nussbaum (1990). The approach has also been used in development studies and quality of life studies (see: Robeyns (2005), p. 93;. Robeyns (2006), p. 351. The United Nations Development Programme (UNDP) has also used the capability approach to rank countries in a Human Development Index, which is an alternative to standard GDP rankings.

[85] Nussbaum (2000), n 81, 5.

[86] Ibid at 71.

[87] Nussbaum (2000), n 81.

[88] Sen (2009), n 78, 286.

[89] Anand and van Hees (2006), p. 268; Anand et al. (2009), p. 125.

capabilities include the capability to live to the end of human life of normal length, have good health, and live with and concern for and in relation to animals, plants, and the world of nature.[90] These central capabilities seem in line with the sustainability-related objectives stipulated by the EU legislature.[91]

One of the main objections to the capability approach is the possibility that different people(s) may place different weights on the capabilities. To deal with this problem, academics have shown that the capabilities proposed by Nussbaum can be supported by empirical evidence derived from data on preferences and happiness.[92] As such, capabilities can be measured. Such measurements allow capabilities to be weighed with preferences, happiness, and resources in a quantitative form based on the model of a CBA. I will discuss this weighing act further in the following section.

5.5 Incorporating HWB Analysis in Article 101(3) TFEU

To conduct a CBA in terms of HWB, clarity is needed regarding the choice of a discount rate, the choice of what values society deems important (capabilities), and how the outcome of the analysis should be weighed and balanced. It should be noted that these questions are not unique to the HWB analysis, as similar questions need to be answered as under any other method that assesses welfare or well-being. However, for purpose of this chapter, I will focus more specifically on providing clarity regarding these issues with respect to the HWB analysis.

5.5.1 Choice of a Discount Rate

As the choice of a discount rate, i.e. the value society puts on future benefits over present benefits, has far-reaching consequences on the outcome of any analysis that assesses interventions that may affect future generations, it should be determined who has the legitimacy to decide on the choice of a discount rate, and whether using a discount rate is legitimate.[93]

The judgments by the CJEU seem to suggests that the use of a discount rate could be legitimate. In Asnef-Equifar v. Ausbanc the CJEU held that persons who are

[90] Nussbaum (2000), n 81, 78–80.

[91] E.g. Article 3 TEU (well-being of EU peoples); Article 2 Charter of Fundamental Rights (CFR) (right to life), Article 4 CFR (prohibition of torture); Article 35 CFR (health care); Article 37 CFR (environmental protection); Article 13 TFEU (animals are sentient beings); Article 11 TFEU (high level of environmental protection).

[92] Anand et al. (2009), n 81.

[93] Some scholars even question the legitimacy of discounting altogether. Philibert (1999), p. 913.

unable to obtain a service (getting loans) as a result of an anti-competitive agreement (disclosing credit history of potential customers of banks) obtained a benefit (avoid over-indebtedness).[94] As such, "the beneficial nature of the effect on all consumers in the relevant markets that must be taken into consideration, not the effect on each member of that category of consumers."[95] This could be interpreted as that the beneficial nature of consumers in current and future markets must be considered. Such an interpretation is strengthened by the CJEU in GlaxoSmithKline where the CJEU held that a restriction on competition today could be exempted when it would provide future benefits.[96] In addition, the CJEU argued in MasterCard that when restriction of competition is found on one side of the market, some of the advantages must also fall on that side of the market.[97] Applying this rationale to an agreement that affects present and future markets, restrictions on competition in present markets providing advantages to future markets can only be exempted when present markets have some advantages by the agreement. As such, the application of a discount rate to future benefits may be warranted: not applying a discount rate to future benefits would imply the impoverishment of current generations.[98]

Although the judgment of the CJEU suggests that the use of a discount rate could be legitimate, the CJEU is not clear on which discount rate to apply, and whether either a constant discount rate should be applied (i.e. all future costs and benefits are discounted at the same rate) or a time declining discount rate (i.e. the discount rate for future costs and benefits declines by time). The Guidelines to Article 101(3) TFEU, also do not specify the choice of a discount rate.[99] However, in the Better Regulation guidelines, the Commission recommends the use of a constant discount rate of 4% for short time frames.[100] The Commission finds it appropriate to have a time declining discount rate for longer time frames.[101] In its toolbox to better regulation, the Commission gives the UK government as an example, which uses a 3.5% discount rate for periods up to 50 years which declines to 1.0% when the time horizon exceeds 300 years.[102] The choice of a discount rate is a highly political matter as it reflects society's preferences between present and future consumption. For that reason, the choice of a discount rate should be left to the individual Member States. To allow undertakings to conduct an HWB assessment of their agreements under Article 101 TFEU, Member States should clarify which discount rate undertakings should apply.

---

[94] Case 238/05 *ASNEF-EQUIFAX*, ECLI:EU:C:2006:734.

[95] Ibid, at 70.

[96] Case 501/06P *GlaxoSmithKline*, ECLI:EU:C:2009:610.

[97] Case 382/12 P *MasterCard*, ECLI:EU:C:2014:2201, at 242; Monti and Mulder (2017), n 4, 650.

[98] Olson and Bailey (1981), n 67.

[99] Article 81(3) Guidelines, n 3, at 88.

[100] Better Regulation, n 47.

[101] Ibid.

[102] Hall et al. (2010).

5.5.2 *Choice of Capabilities*

To include capabilities in the CBA, it should be determined who has the legitimacy to decide which capabilities are important. Although those who may be quick to point out that a similar decision does not need to be made under a welfare economics approach, fail to acknowledge that a welfare economic approach is also far from neutral or free from moral considerations as it is based on the idea that the only thing of value in life is preference satisfaction.

To decide on the capabilities, Martha Nussbaum has provided an elaborate and extensive list of which abilities or opportunities are necessary to enable individuals to function in a specific way.[103] This list could be used as guidelines to conduct the HWB analysis, leaving room for Member States to deviate from the list, if they so desire, allowing the list to be part of a democratic process.[104]

5.5.3 *Weighing and Balancing*

The HWB analysis produces separate indicators on preferences, happiness, resources, and capabilities. These indicators will have to be weighed and balanced against each other to allow a decision to be made on whether the sustainability agreement benefits the consumer. The practice of weighing and balancing is already quite common in law where different norms and practices are often balanced to reach acceptable outcomes.[105]

When the legitimacy to conduct such a balancing exercise is not delegated to undertakings,[106] the authority that has the legitimacy to weigh and balance the separate indicators should construct composite indicators. Composite indicators combine individual indicators with different units of measurement into a single measure.[107] To produce a composite indicator, component indicators must be selected and weighted. As such, the information provided by composite indicators varies according to the weighing and balancing used for their construction making them more susceptible to manipulation.[108] For that reason, composite indicators must be constructed by the legitimate authority with appropriate checks and balances.

---

[103] Nussbaum (2000), n 81.

[104] As suggested by Sen (2009), n 78.

[105] Gerbrandy (2017), n 4.

[106] Claassen and Gerbrandy argue that such legitimacy could also be delegated to undertakings, Claassen and Gerbrandy (2018), p. 401.

[107] Eurostat, 'Towards a harmonized methodology for statistical indicators. Part 1: Indicator typologies and terminologies' (2014).

[108] Ibid.

5.6 Conclusion

This chapter starts from the premise that further guidance is needed to allow undertakings to conduct an analysis under Article 101(3) TFEU of sustainability agreements that yield benefits that have no clear market price. It is argued that, based on communications by the Commission as well as CJEU jurisprudence, benefits under Article 101 TFEU should be estimated by taking a wider notion of well-being into account that goes beyond purely monetary well-being. This chapter proposes the method of HWB to estimate the benefits of an agreement under Article 101(3) TFEU in a way that is quantifiable but goes beyond the economic approach focusing solely on monetary well-being. In addition, it is argued that when EU or national authorities want to exempt agreements aimed at achieving a public policy under Article 101(1) TFEU, best practices demand that they make an accurate estimation of the impact the agreement has on a wider notion of well-being.

This chapter argues that to allow undertakings to assess whether their agreement fulfils the conditions of Article 101 TFEU, the relevant legitimate authority should answer questions regarding the choice of a discount rate, the choice of the capabilities, and how the separate indicators of the HWB analysis should be weighed and balanced. When the legitimacy to conduct such a balancing exercise is not delegated to undertakings, it is suggested that the authority that has the legitimacy to weigh and balance the separate indicators should combine the individual indicators with different units of measurement into a single measure (also known as composite indicators) that can then be used by undertakings to assess their agreement.

In sum, this chapter shows the value of HWB indicators as (1) an addition to the information base of the legal authority and (2) a guidance to undertakings in their self-assessments under the competition rules. Using these indicators could result in a more accurate and transparent analysis under Article 101 TFEU. In the right hands, with the proper understanding of their strengths and weaknesses, the HWB indicators can provide important insights into what extent a private sustainability initiative can be justified or exempted from Article 101 TFEU. As such, HWB analysis can provide otherwise unattainable insights on the benefits of a private sustainability initiative and provide appropriate guidance to undertakings in their self-assessment of their sustainability initiatives under the EU competition rules, thereby stimulating private initiatives that further sustainable consumption. Misused or misunderstood, HWB analysis, like any other type of analysis, can be misleading and result in unjustified outcomes.

References

Adler MD, Posner EA (2006) New foundations of cost-benefit analysis. Harvard University Press
Adler M, Posner E (2008) Happiness research and cost-benefit analysis. J Legal Stud 37:253
Alexy R (2003) Constitutional rights, balancing, and rationality. Ratio Juris 16:131

Anand P, van Hees M (2006) Capabilities and achievements: an empirical study. J Soc-Econ 35(2): 268–284

Anand P et al (2009) The development of capability indicators. J Human Dev Capabil 10:125

Anderson E (2010) Justifying the capabilities approach to justice. In: Brighouse H, Robeyns I (eds) Measuring justice primary goods and capabilities. Cambridge University Press

Beck U (2014) The brave new world of work. John Wiley & Sons

Beck U (1992) Risk society. Towards a new modernity. Sage Publications

Bentham J (2007) An introduction to the principles of morals and legislation. Dover Publications Inc

Bronsteen J, Buccafusco C, Masur J (2013) Well-being analysis vs. Cost-benefit analysis. Duke Law J 62:1603

Browning E, Zupan M (2009) Microeconomics: theory & applications. John Wiley & Sons

Buchanan J (1987) Opportunity cost. The New Palgrave: a dictionary of economics, 3rd edn, pp 718–721

Casey D (2009) Disintegration: environmental protection and Article 81 EC. Eur Law J 15:362

Claassen R, Gerbrandy A (2016) European competition law: from a consumer welfare to a capability approach. Utrecht Law Rev 12(1):5

Claassen R, Gerbrandy A (2018) Doing good together: competition law and the political legitimacy of interfirm cooperation. Bus Ethics Q 28:401

Daskalova V (2015) Consumer welfare in EU competition law: what is it (not) about? Compet Law Rev 11:131–160

De Burcá G (1993) The principles of proportionality and its application in EC law. Yearb Eur Law 13:105

DeGraba P (2006) The loss leader is a Turkey: targeted discounts from multi-product competitors. Int J Ind Organ 24:613–628

Dolmans M (2020) Sustainable competition policy. Compet Law Policy Debate 6(1):17

Feldstein M (1964) The social time preference discount rate in cost benefit analysis. Econ J 74(294): 360–379

Frederick S, Loewenstein G (1999) Hedonic adaptation. In: Kahneman D, Diener E, Schwarz N (eds) Foundations of hedonic psychology: scientific perspectives on enjoyment and suffering. Russell Sage Foundation, p 302

Gerbrandy A (2013) Competition law and private-sector sustainability initiatives. In: Ciacchi AC et al (eds) Law & governance - beyond the public-private divide? Eleven International Publishing

Gerbrandy A (2017) Solving a sustainability-deficit in European competition law. World Compet 40:539

Gilbert D, Wilson T (2007) Prospection: experiencing the future. Science 317:1351

Glen D, others (eds) (2012) Transformative consumer research for personal and horizontal collective well-being. Routledge, p 257

Gollier C (2002) Time horizon and the discount rate. J Econ Theory 107(2):463–473

Hall J et al (2010) A framework to measure the progress of societies. ECD Statistics Working Papers, vol 5. OECD Publishing

Hardin G (1968) The tragedy of the commons science. Science 162:1243

Jan Ankersmit L (2017) Green Trade and Fair Trade in and with the EU. Process-based measures within the EU legal order. Cambridge University Press

Jansen C, Kloosterhuis E (2016) The wouters case law, special for a different reason. Eur Compet Law Rev 37:335

Kahneman D, Diener E, Schwarz N (1999) Well-being: the foundations of hedonic psychology. Russell Sage Foundation

Keating B, Keating M (2017) Basic cost benefit analysis for assessing local public projects, 2nd edn. Business Expert Press

Kingston S (2012) Greening EU competition law and policy. Cambridge University Press

Klatt M, Meister M (2012) The constitutional structure of proportionality. Oxford University Press

Luhmann M et al (2012) Subjective well-being and adaptation to life events: a meta-analysis. J Pers Soc Psychol 102:592
Monti G (2002) Article 81 EC and Public Policy. Common Mark Law Rev 39:1057
Monti G (2020) Four options for a greener competition law. J Eur Compet Law Pract
Monti G, Mulder J (2017) Escaping the clutches of EU competition law: pathways to assess private sustainability initiatives. Eur Law Rev 42:641
Mortelmans K (2001) Towards convergence in the application of the rules on free movement and competition? Common Mark Law Rev 38:613
Nas T (2016) Cost-benefit analysis: theory and application. Lexington Books
Norgaard R, Howarth R (1991) Sustainability and discounting the future. Ecological economics: the science and management of sustainability. Columbia University Press, pp 88–101
Nowag J (2016) environmental integration in competition and free-movement laws. Oxford University Press
Nussbaum M (1988) Nature, function, and capability: aristotle on political distribution. Oxf Stud Anc Philos:145
Nussbaum M (1990) Aristotelian Social Democracy. In: Bruce Douglass R, Mara GM, Richardson HS (eds) Liberalism and the Good. Routledge
Nussbaum M (2000) Woman and human development: the capabilities approach. Cambridge University Press
Odudu O (2006) The boundaries of EC competition law: the scope of article 81. Oxford University Press
Odudu O (2010) The wider concerns of competition law. Oxf J Leg Stud 30:599
Olson M, Bailey M (1981) Positive time preference. J Polit Econ 89:1
Padilla E (2002) Intergenerational equity and sustainability. Ecol Econ 41(1):69–83
Philibert C (1999) The economics of climate change and the theory of discounting. Energy Policy 27:913
Platt J (1973) Social traps. Am Psychol 28:641
Pogge T (2005) World poverty and human rights. Ethics Int Aff 19:1
Rawls J (1982) Primary goods. In: Sen AK, Williams BAO (eds) Utilitarianism and beyond. Cambridge Univesrity Press, pp 86–87
Robeyns I (2005) The capability approach: a theoretical survey. J Hum Dev 6:93
Robeyns I (2006) The capability approach in practice. J Polit Philos 14:351
Schelling T (1984) Self-command in practice, in policy, and in a theory of rational choice. Am Econ Rev 74:1
Sen A (1987) The Standard of Living. Cambridge University Press, p 17
Sen A (2009) The idea of justice. Harvard University Press, p 286
Sen AK (1999) Commodities and Capabilities. Oxford University Press
Soanes C, Stevenson A (2005) Oxford Dictionary of English. Oxford University Press
Stiglitz J, Sen A, Fitoussi J-P (2009) The measurement of economic performance and social progress revisited. OFCE, p 33
Sunstein CR (2018) The cost-benefit revolution. MIT Press
Townley C (2009) Article 81 EC and public policy. Hart
Townley C (2011) Which goals count in Article 101 TFEU? Public policy and its discontents: the OFT's roundtable discussion on article 101(3) of the TFEU. Eur Compet Law Rev 32:441
Van den Bergh RJ (2017) Comparative competition law and economics. Edward Elgar Publishing
Whish R, Bailey D (2015) Competition law, 8th edn. Oxford University Press, p 140
Wiener J (2006) Better regulation in Europe. Curr Leg Probl 59:447–518
Wilson T, Gilbert D (2005) Affective forecasting: knowing what to want. Curr Dir Psychol Sci 14: 131

Chapter 6
Conclusion and Discussion

The presented chapters taken together show that the extent to which the interpretation and application of EU law to sustainability labels inhibits sustainable consumption depends on the specific normative assumptions underlying legal interpretation. Using critical legal analysis as a theoretical framework, it was found that most commonly, the interpretation and application of EU law uses normative assumptions derived from a noncritical application of neoclassical law and economics. In this book, I went through three legal frameworks and showed that when these normative assumptions on how consumers perceive sustainability attributes on food products in each of these frameworks changed, the inhibitive effect of the respective legal framework on sustainable consumption changed as well. Results show that the use of normative assumptions based on a noncritical application of neoclassical law and economics has an inhibiting effect on sustainability labels that promote best practices in sustainable development (e.g. sustainability labels that promote human well-being are prohibited as benefits are solely assessed in monetary terms), while it leaves more space to sustainability labels that do not promote such best practices (e.g. sustainability labels that greenwash products are not prohibited because the consumer is considered to easily distinguish greenwashed products from sustainable products).

This book shows that within the examined legal frameworks, the critical application of insights from the social sciences results in either (1) more regulatory space for sustainability labels that promote best practices in sustainable development without relaxing constraints on sustainability labels that do not, or (2) less regulatory space for greenwashing sustainability labels without prohibiting sustainability labels that promote best practice. Regulatory space is an analytical construct employed to describe and examine the environment within which regulation takes place.[1] Within

[1] Windholz (2017) at Chap. 4; See Hancher and Moran (1989), who are generally acknowledged for having coined the term, Hancher and Moran (1989). See Scott (2001) about a more elaborate discussion on the concept of regulatory space and the role of law therein, Scott (2001), p. 329.

© The Author(s), under exclusive license to Springer Nature Switzerland AG 2022
E. van der Zee, *Sustainability Labels in the Shadow of the Law*, Studies in European Economic Law and Regulation 22, https://doi.org/10.1007/978-3-030-95802-2_6

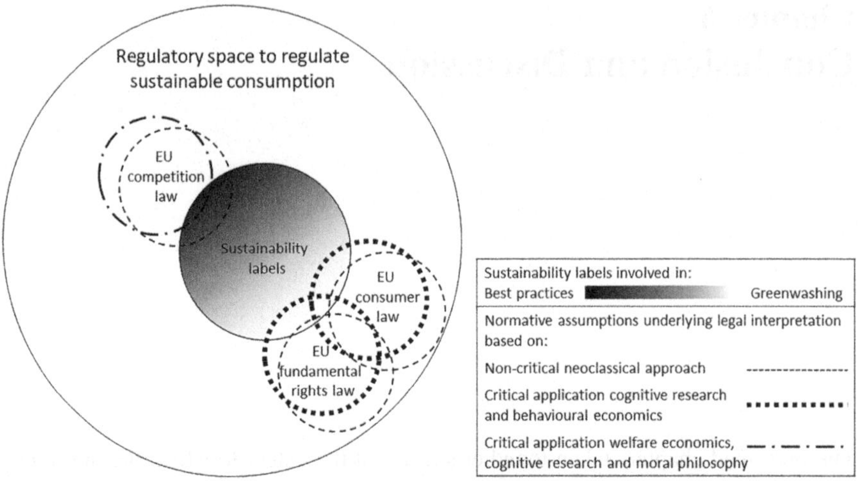

Fig. 6.1 The increase and decrease of the regulatory space of sustainability labels depending on the normative assumptions underlying legal interpretation per legal framework

this regulatory space there are exchanges and interdependence relationships between a range of non-state and state organizations, which are linked through networks and often compete for power.[2] These exchanges and interdependence relationships may be governed by law. Laws, therefore, also exist in the regulatory space,[3] and may compete for power with sustainability labels. Figure 6.1. illustrates the increase and decrease of the regulatory space of sustainability labels depending on the normative assumptions underlying legal interpretation.

As such, it was found that when in the interpretation and application of EU consumer law to sustainability labels the underlying normative assumptions are based on a non-critical neoclassical approach, only a fraction of the sustainability labels that are involved in greenwashing practices would be prohibited (such as sustainability labels that use the textual term "organic" while they do not comply with the EC Regulation on Organic Agriculture[4]). Most greenwashing sustainability labels (such as those using visual information that consumers may associate with increased sustainable production) would not be prohibited. It was found that when these normative assumptions are based on a critical application of cognitive science and behavioural economics, more greenwashing sustainability labels would be prohibited (most notably those sustainability labels that use visual information that consumers may associate with increased sustainable production).

This book shows that when in the interpretation and application of EU fundamental rights law the underlying normative assumptions are based on a noncritical

[2] Lange (2003).

[3] Ibid.

[4] EC Regulation on Organic Agriculture, Chap. 2, n 40, Articles 23 and 25.

application of neoclassical law and economics, public authorities have less leeway to reduce the regulatory space of sustainability labels. The reason for this is that they have fewer possibilities to ban potentially misleading information on food products as disclaimers are assumed to have the same effectiveness as a complete ban[5] to achieve a public policy objective (such as consumer or environmental protection). Insights from cognitive science and behavioural economics show that information overload makes it more difficult for consumers to process information to make an informed choice of sufficient quality.[6] When information overload is apparent, a complete ban may be more effective. As such, using insights from cognitive science and behavioural economics in the interpretation and application of EU fundamental rights law may call for a prohibition of products instead of dislcaimers when cognitive research shows that it is more effective to achieve public policy objectives (such as sustainable consumption).

It was also found that, when critically applied, science-based estimations could be an addition to the information base of the legal authority and a guidance to undertakings in their self-assessment under the EU competition rules. As such, critically applying insights from welfare economics by taking into account insights from cognitive science and moral philosophy in the interpretation and application of EU competition law would stimulate private actors to initiate agreements based on best practices as these would not be prohibited by EU competition law. As such, these agreements have the potential to reduce the number of unsustainable products, as well as greenwashing labels on the market, thereby improving sustainable consumption. In the same vein, critically applying these insights guides the legal authority to more accurately assess whether agreements promote best practices in sustainable production.

In the next section (Sect. 6.1) the answer to sub-question 1 is discussed, which provides clarity as to which sustainability labels are susceptible to greenwashing and which sustainability labels are more likely to promote best practices in sustainable development. In Sect. 6.2., the answer to sub-questions 2 to 4 is discussed, which provides clarity as to how different normative assumptions underlying legal interpretation and application of EU consumer law, EU fundamental rights law, and EU competition law changes the boundaries of the regulatory space of sustainability to promote sustainable consumption. In Sect. 6.3. the limitations of this book and the contributions of this book are discussed.

---

[5]This was the case in Case 120/78 *Cassis de Dijon*, ECLI:EU:C:1979:42; Case 178/84 *Beer Purity*, ECLI:EU:C:1987:126. See on this information paradigm: Purnhagen (2014), Chap. 4, n 60.
[6]Verbeke (2005), Chap. 4, n 158.

6.1 The Regulatory Structure of Sustainability Labels

This section will answer the first sub-question of this book, i.e. what is the regulatory structure of sustainability labels? It was shown in Chap. 2 that mainly private actors are involved in creating sustainability labels. Three structural patterns of sustainability labels were identified on a spectrum ranging from multi-stakeholder regulatory instruments to self-regulatory instruments.

Sustainability labels taking the form of multi-stakeholder regulatory instruments appear most frequently.[7] The regulatory structure of these sustainability labels includes high NGO involvement in collaboration with firms in standard-setting, and certification conducted by private third parties. Multi-stakeholder regulatory instruments bring stakeholders together to participate in standard-setting making it easier for the producers involved to voice concerns with specific standards, giving them some influence in standard-setting. Furthermore, sustainability labels based on multi-stakeholder regulatory instruments resemble a structure that is trusted most by consumers and considered most reliable and credible.[8]

It seems that these multi-stakeholder regulatory instruments illustrate best practices for any regulatory instrument to promote sustainable development.[9] However, an important pitfall of multi-stakeholder regulatory instruments is that they are susceptible to asymmetry in stakeholder involvement which may lead to knowledge gaps[10] and a lack of representativeness.[11] Furthermore, as many multi-stakeholder regulatory instruments operate internationally they may encounter cultural and institutional differences that may make stakeholder involvement difficult.[12] Finally, as stakeholders may be involved in different ways (by having voting rights, or as consultants, advisors or co-regulators), there is a risk that stakeholders may be neglected in the decision-making phase.[13]

The second structural pattern of sustainability labels that was found is characterized by self-regulatory instruments. Within these types of sustainability labels, labels are created solely by firms, and certification is conducted through private first-party certification, meaning that the firm certifies its own products. In some cases, these self-regulatory instruments require that the product complies with multi-stakeholder

---

[7] These are also the standards that are frequently discussed by legal and governance scholars. See e.g.Bernstein and Hannah (2008), p. 575; Cafaggi (2011), p. 20.

[8] Golan et al. (2001); Hatanaka et al. (2005), Chap. 2, n 43; Padel and Foster (2005), Chap. 2, n 26; Eden et al. (2008), Chap. 2, n 25.

[9] Some multi-stakeholder regulatory instruments have even been used by public authorities to procure fair trade products that bear the private label awarded when product complies with such a multi-stakeholder regulatory instrument. In the case of Max Havelaar, the ECJ considered such practice to violate the EU public procurement directive (directive 2004/18/EC). See Case 368/10 *Commission v the Netherlands*, ECLI:EU:C:2012:284.

[10] Gerrits and Edelenbos (2004), p. 239.

[11] Marx et al. (2012), p. 90; Cheyns (2011).

[12] Sperber (1996); Mitchell (2002).

[13] Gerrits and Edelenbos (2004), n 10.

regulatory instruments, in other cases these instruments focus on adopting unilateral policies that apply to their operations or purchases. These unilateral policies are generally not publicly available.[14] As these claims are not verified by external auditors, nor are the policies underlying these claims publicly available, consumers and investors have to rely on the company's reputation in evaluating the sustainability impact of production.[15] For that reason, such sustainability labels are more susceptible to greenwashing than multi-stakeholder regulatory instruments, meaning that the firm may commit to its unilateral policies to promote sustainable development while not implementing them.[16]

In between these market-based regulatory instruments and the self-regulatory instruments, a third structural pattern was identified. These types of labels are characterized by sector-specific regulatory instruments. Within these types of sustainability labels, labels are created by NGOs within a specific sector without collaboration with firms and require second-party certification. These sustainability labels are based on established guidelines that their corporate members have to comply with to make specific claims. Similar to the self-regulatory instruments, the guidelines are generally not publicly available. As these claims are verified by the NGO that created the sustainability label and not by a third party, and because the guidelines underlying these claims are generally not publicly available, consumers and investors have to rely on the NGO's reputation in evaluating the sustainability impact of production. As such, these sector-specific regulatory instruments are also more susceptible to greenwashing.[17]

The existence of a multiplicity and diversity of sustainability labels with different degrees of reliability, credibility and effectiveness on the consumer market may reduce the potential of all sustainability labels to promote sustainable development. Cognitive research shows that when specific cues, which may include pictograms on food packaging used by sustainability labels, are valued highly by consumers, the behaviour of consumers seems largely controlled by the associations that come to mind most easily.[18] Such associations may be overridden by a more deliberate mode of operation,[19] but such a mode of operation may not appear in a supermarket setting where shopping is often a hurried affair where different information cues compete for attention.[20] The multiplicity and diversity of pictograms on the market that consumers may associate with sustainable production make it very difficult for

[14]See also Vandenbergh (2007), p. 913.

[15]Lipschutz and Fogel (2002), p. 134.

[16]See e.g. Ramus and Montiel (2004), p. 377; Winn and Angell (2000), p. 1119.

[17]See e.g. Ramus and Montiel (2004); Winn and Angell (2000).

[18]This process is known as peripheral processing, heuristic processing, system 1 processing or impulsive system. See e.g. Kahneman (2003), p. 697; Hoogland et al. (2007), Chap. 2, n 56, 47; Strack and Deutsch (2004), p. 220; Chernev and Carpenter (2001), p. 349; Purnhagen et al. (2016).

[19]This process is known as central processing, systematic processing, system 2 processing or reflective system. See ibid.

[20]Horne, Chap. 2, n 5.

consumers to know and understand every single sustainability label. As a result, when consumers have positive associations with a pictogram used by a sustainability label, due to the credibility and reliability of a similar pictogram used by another sustainability label, these associations may positively affect consumer perception towards sustainability labels that use similar pictograms but that are mere examples of greenwashing. As sustainable production and independent checks to guarantee sustainable production may warrant a price premium to consumers,[21] food businesses may be incentivized to free-ride on those pictograms with which consumers have positive associations. This may potentially lead to a loss of customers for the reliable and credible sustainability labels, reducing their potential to promote sustainable consumption.

The potential of sustainability labels to promote sustainable consumption depends, thus, not only on the extent to which the sustainability label is effective in promoting sustainable production processes due to the manipulation of global markets,[22] but depends also on how the information used by the sustainability label is processed by consumers. This latter effect was taken into account within this book to investigate the interaction between the interpretation and application of EU law and the regulatory space of sustainability labels to promote sustainable development.

6.2 Interaction Between the Interpretation and Application of EU Law and Sustainability Labels

In this section sub-question 2, 3, and 4 are discussed. The discussion will be based upon a comparison of findings across the three different case studies as presented in Chaps. 3, 4 and 5. This discussion provides more clarity as to how different normative assumptions underlying legal interpretation and application of EU law changes the boundaries of the regulatory space of sustainability labels to promote sustainable development. Through analysing the four cases, this book examined the various ways in which EU law interacts with sustainability labels in more detail through specific case study analysis. Within the case study analysis, the interaction between four sets of legal frameworks (EU consumer law, EU fundamental rights law, and EU competition law) and sustainability labels was examined. These legal frameworks were selected to examine when sustainability labels fall within the scope of EU law (sub-question 2); to what extent EU law leaves room to interpret and apply legal provisions to sustainability labels at the national and EU level (sub-question 3); and whether and to what extent findings from other scientific disciplines could inform the doctrinal analysis of the interpretation and application of EU law to sustainability labels (sub-question 4). Table 6.1 provides a brief

[21] Cafaggi and Iamiceli (2013), Chap. 3, n 2.

[22] Cashore (2002).

Table 6.1 Findings of the book

Case study		EU consumer law (Chapter 3)	EU fundamental rights law (Chapter 4)	EU competition law (Chapter 5)
Legal provision		Article 7 FIR	Article 11 CFR	Article 101 TFEU
Sustainability label in scope of legal framework (sub-question 2)		Sustainability label provides food information to final consumer	Expression used by sustainability label is limited	Sustainability label is an agreement between undertakings engaged in economic activity aimed to make a profit
Interpretation and application of legal provisions (sub-question 3)	National level	Decide whether sustainability labels give misleading impression *in concreto;* Sustainability label using textual information indicating "organic" while product is not in compliance with EU Regulation on Organic Agriculture should always be prohibited	Decide whether limitation is proportional: • Necessity test (least onerous option of the available options) - Disclaimers preferred when same effectiveness as complete ban - Public health trumps free speech as a default • And regulation must meet objectives of general interest recognized by the EU or regulation protects rights and freedoms of others	When sustainability label affects trade between Members, conduct proportionality test in which it should be assessed whether the sustainability label ensures human well-being. NB: if NCAs find that the sustainability label ensure HWB they can only state there are no grounds for action on their part
	EU level	Provides guidance of what can be expected of average consumer when confronted with textual information; Provides regulation which commercial practices are considered unfair in all circumstances	Ibid.	When sustainability label affects trade between Members, conduct proportionality test in which it should be assessed whether the sustainability ensures human well-being. NB: When Commission acts on own initiative it can bring infringement to an end; exempt sustainability label from competition scrutiny; justify sustainability label

(continued)

Table 6.1 (continued)

Case study	EU consumer law (Chapter 3)	EU fundamental rights law (Chapter 4)	EU competition law (Chapter 5)
Additional evidence of relevance (sub-question 4)	Cognitive and behavioural economics studies into what is misleading to actual consumers	Cognitive and behavioural economics studies into the effectiveness of EU Regulation limiting commercial expression to protect public interest objective (e.g. sustainable consumption)	Social and ethical considerations on estimating human well-being

overview of the findings of this book, which will be further elaborated upon in Sects. 6.2.1, 6.2.2, and 6.2.3.

6.2.1 Sustainability Labels Within the Scope of EU Law

This section answers and discusses sub-question 2: When do sustainability labels fall within the scope of EU law? This book shows that whether sustainability labels fall within the scope of the examined legal framework depends on the specific legal framework at issue. Under the Food Information Regulation (FIR),[23] sustainability labels fall within its scope when it provides information concerning a food product to the final consumer by any means, e.g. on the food product itself or the company website.[24] Under Article 11 EU Charter of Fundamental Rights (CFR), sustainability labels fall within its scope when public authorities limit information concerning a food product to the final consumer. Sustainability labels fall within the scope of EU competition law when the sustainability label is an agreement between undertakings engaged in economic activity aimed to make a profit that has at its object or effect to prevent, restrict or distort competition within the internal market.

[23] EU Food Information Regulation, Chap. 1, n 57.

[24] Article 2(2)(a) FIR.

6.2.2 *Interpretation and Application of Legal Provisions Relevant to Sustainability Labels That Fall within Its Scope at National and EU Level*

This section answers and discusses sub-question 3: To what extent does EU law leave room for interpreting and applying legal provisions to sustainability labels at the national and EU level? This book shows that the extent to which EU law leaves room to interpret and apply legal provisions to sustainability labels depends on the legal interpretation of the proportionality of the trade restriction by the sustainability labels to achieve a sustainability objective. The exercise of this proportionality test varies from case to case, from jurisdiction to jurisdiction and according to which goal it is applied.[25] In this book, the proportionality test was discussed in the context of EU consumer law, EU fundamental rights law, and EU competition law.

In the context of EU consumer law, the room for interpretation and application lies within the margin of discretion left by the ECJ to national authorities in the legal assessment of whether commercial information is misleading *in concreto*. This margin of discretion follows from the proportionality principle which requires that actions by the EU and its Member States is proportionate, e.g. the action must (1) be a useful, suitable, or effective means of achieving a legitimate aim; (2) not be more restrictive than necessary to achieve that legitimate objective; (3) not have an excessive or disproportionate effect on the other interests.[26] Secondary legislation provides further regulation as to how national authorities should assess commercial information. Most relevant to this book is that when companies use textual information such as “organic”, “eco”, and “bio” or the EU pictogram for organic agriculture on their food packaging while the food product does not comply with the EU Regulation on Organic Agriculture the margin of discretion left to the national authorities to assess whether such information is misleading is zero: these types of labels must be prohibited in all EU Member States.[27] That would also be the case when a company claims to be a signatory to a code of conduct while this is not the case, displays other sustainability pictograms without having obtained the necessary authorisation, claims that a code of conduct has an endorsement from a public or other body which it does not have, or claims that its product has been approved, endorsed or authorised by a public or private body while this is not the case.[28]

In other circumstances, the margin of discretion left to national courts to decide whether commercial information is misleading *in concreto* depends on the legal interpretation and application of the normative “average consumer” benchmark by

[25] See e g for EU Law: Harbo (2010); also Tridimas (1999).

[26] De Burcá (1993), Chap. 5, n 31.

[27] EC Regulation on Organic Agriculture, Chap. 2, n 40, Articles 23 and 25.

[28] Annex 1(1)-(4) UCPD.

the ECJ.[29] To guide national courts in the legal assessment of misleading information, the ECJ developed the normative "average consumer" benchmark as an expression of the proportionality test with regard to the interpretation of the free movement of goods,[30] which has later been taken up by secondary legislation.[31] Under this benchmark, the "average consumer" is considered "reasonably well informed and reasonably observant and circumspect", "taking into account social, cultural and linguistic factors".[32] Hence, for national courts to conduct a legal assessment of misleading information under the FIR, they have to interpret and apply the "average consumer" benchmark. This book shows that when interpreting and applying the "average consumer" benchmark, national courts could use insights from cognitive science and behavioural economics. Such an understanding is in line with the general trend in the case-law of the ECJ to leave open the possibility for Member States to use statistical evidence or any other suitable type of evidence to justify measures infringing the free movement of goods, where the "average consumer" benchmark is part of.[33] Furthermore, as the Lisbon Treaty has reinforced the consumer dimension of EU internal market law,[34] such insights from cognitive science and behavioural economics provide a better understanding of how consumers process information.

In the context of EU fundamental rights law, the room for interpretation and application lies within the margin of discretion left by the ECJ to EU and national courts in the legal assessment of whether a limitation to the fundamental right to freedom of expression on food labels, which is often the result of the application of EU secondary legislation,[35] is proportional. This requires a judicial review as to whether the EU secondary legislation hinders the exercise of the freedom of expression only to the extent necessary to achieve the public policy objective. This book shows that the width of the margin of discretion in the legal assessment of limitations

[29] See further on the margin of discretion left to national courts under the UCPD by the ECJ, Stuyck (2015).

[30] The ECJ has consistently used this wording since its judgment of 16 July 1998, Case 210/96 *Gut Springheide*, ECLI:EU:C:1998:369, at 37. Prior to this decision the Court had already referred to the "[r]easonably circumspect consumer" as yardstick, Case 470/93 *Mars*, ECLI:EU:C:1995:224, Chap. 3, n.5, at 13: Franck and Purnhagen (2014), p. 336.

[31] Such as EU Food Information Regulation, Chap. 1, n 57.

[32] See e.g., Case 315/92 *Clinique*, ECLI:EU:C:1994:34 and Case 210/96 *Gut Springheide*, ECLI:EU:C:1998:369.

[33] Case 333/14 *Scotch Whiskey*, ECLI:EU:C:2015:845, Chap. 3, n 45, at 64-5 in conjunction with Case 148/15 *Deutsche Parkinson Vereinigung*, ECLI:EUC:2016:776, Chap. 3, n 45, at 35-6.

[34] See e.g. Article 12 TFEU and Article 38 CFR. See also De Vries (2012), pp. 228, 236–238.

[35] Within the European Union many legally binding rules, mainly Regulations but also Directives, relate to information provided for on food packaging. Examples are EU Food Information Regulation, Chap. 1, n 57; EC Regulation on Organic Agriculture, Chap. 2, n 40;

EC Regulation on genetically modified food and feed, Chap. 4, n 46; EC Regulation traceability and labelling of genetically modified organisms, Chap. 4, n 46; EC Regulation on the nutrition and health claims made on foods, Chap. 4, n 46; Regulation regarding the labelling of beef and beef products, Chap. 4, n 46.

set by the EU is broader for commercial expression than political expression. Textual and visual information used by sustainability labels can only be considered commercial in nature under Article 11 CFR and not political. Therefore, the EU legislator has a wide margin of discretion with regard to the limits set to commercial expression (such as those used by sustainability labels), especially when the public policy objective the limitation pursues is the protection of public health.[36]

In the context of EU competition law, the room for interpretation and application lies within the margin of discretion left by the ECJ and national courts to economic regulators in the legal assessment of the proportionality of a private sustainability agreement under Article 101 TFEU. When public authorities (either at EU or national level) would want to exempt a private sustainability agreement from competition scrutiny following the *Wouters* doctrine, a proportionality test needs to be conducted. To justify an anti-competitive private sustainability agreement under Article 101(3) TFEU, a proportionality test should also be conducted. This book argues that to conduct the proportionality test under the *Wouters* doctrine and 101(3) TFEU, insights on the impact of the private sustainability agreement on EU human well-being should be taken into account for two reasons. First, the overall aim of the European Union is to promote the well-being of its peoples (Article 3(1) TEU). Second, the ECJ rejects an overly economic notion of consumer welfare as the objective of competition rules, instead favouring that competition rules should improve a wider notion of well-being in the EU.[37] In both instances, the type of "well-being" referred to includes the two components of human well-being: standard of living and quality of life. As such, human well-being includes objective factors (such as living conditions, income, health) and the subjective perception one has of them.[38] Human well-being, thus, does not only take into account economic values but includes socio-technical, cultural and environmental values that will help humans to meet their fundamental needs.[39] Best practices demand that the EU and its Member States willing to exempt private sustainability agreements from competition scrutiny should combine insights from welfare economics, social and economic psychology, and moral philosophy to come to a more accurate estimation of the costs and benefits in terms of human well-being of the private agreement. A

---

[36]Case 249/09 *Novo Nordisk*, ECLI:EU:C:2010:616, Chap. 4, n 21, at 50. He refers to Alexy (2003), Chap. 4, n 53, 440. Translation from Case 249/09 *Novo Nordisk*, ECLI:EU:C:2010:616, Chap. 4, n 21, at 46. The cases are only accessible in French. See *Société de Conception de Presse et d'Edition et Ponson v France*, no 26935/05, ECtHR 6 March 2009; and *Hachette Filipacchi Presse Automobile and Dupuy v France*, app no 13353/05, ECtHR 5 March 2009: "Ainsi, des considérations primordiales de santé publique, sur lesquelles l'Etat et l'Union européenne ont d'ailleurs légiféré, peuvent primer sur des impératifs économiques, et même certains droits fondamentaux comme la liberté d'expression."; Case 376/98 *Germany v Parliament and Council*, ECLI:EU:C:2000:324, Chap. 4, n 27, at 160-1.

[37]Case 501/06P *GlaxoSmithKline*, ECLI:EU:C:2009:610, at 64; Case 52/09 *TeliaSonera Sverige*, ECLI:EU:C:2011:83, 22.

[38]Eurostat, Quality of Life Indicators – Measuring quality of life, <http://ec.europa.eu/eurostat/statistics-explained/index.php/Quality_of_life_indicators_-_measuring_quality_of_life>.

[39]See Smith and Max-Neef (2011).

similar assessment is warranted under Article 101(3) TFEU. The use of any scientific method to estimate the costs and benefits of an agreement requires answers to highly normative and political questions. It is argued that these normative and political questions reflect the difference between the *Wouters* doctrine and Article 101(3) TFEU. The *Wouters* doctrine leaves some margin of discretion to Member States to decide on the proportionality of a private agreement that pursues a legitimate aim solely articulated by the Member State by determining highly normative and political issues (e.g. the choice of a discount rate and/or capabilities. In other cases, i.e. when an undertaking pursues other legitimate aims, the European Commission would have such a margin of discretion.

6.2.3 Findings from Scientific Disciplines Informing the Doctrinal Analysis

This section answers and discusses sub-question 4: whether and to what extent could findings from other scientific disciplines inform the doctrinal analysis of the interpretation and application of EU law to sustainability labels? This book shows that within the margin of discretion to assess the proportionality of the trade restriction of a sustainability label to achieve a sustainability objective there is much room for manoeuvre to include findings from other scientific disciplines to inform the doctrinal analysis of the interpretation and application of EU law to sustainability labels. In EU consumer law, EU fundamental rights law, and EU competition law, Member States are allowed to take the latest scientific insights into account in the legal assessment of the proportionality of their measure.[40] For EU consumer law and EU fundamental rights law insights from cognitive science and behavioural economics could inform the doctrinal analysis as to how consumers process information. Such insights provide further understanding of how the "real" consumer would perceive food information as opposed to how a *homo economicus* would perceive such information. By using insights from cognitive science and behavioural economics, the interpretation and application of the FIR and Article 11 CFR may further the potential of sustainability labels that stimulate best practices in sustainable production by preventing that information on food products used by greenwashing sustainability labels negatively affects consumer perception towards the reliability and credibility of best practice sustainability labels.

In EU competition law, the EU commission and the national competition authorities have a wide margin of discretion to conduct an economic analysis.[41] When a public authority (either EU or national authorities) would exempt private

[40] See in this respect Case 210/96 *Gut Springheide*, ECLI:EU:C:1998:369, at 31; and Case 333/14 *Scotch Whiskey*, ECLI:EU:C:2015:845, Chap. 3, n 45, at 64–65 in conjunction with Case 148/15 *Deutsche Parkinson Vereinigung*, ECLI:EUC:2016:776, Chap. 3, n 45, at 35–36.

[41] Botta and Svetlicinii (2013).

sustainability agreements from competition scrutiny or justify such agreements, best practices demand that the public authority combines insights from welfare economics, cognitive science, and moral philosophy to come to a more accurate estimation of the impact of the agreements on human well-being. These tools could also be used by undertakings in their self-assessment under EU competition rules. When using these insights awareness of the strengths and weaknesses is important, requiring clarification of which approach and method would be most appropriate and when. Critically applying insights derived from welfare economics, cognitive science, and moral philosophy in the interpretation and application of EU competition law may further the potential of sustainability labels to promote sustainable development. Such insights prevent that sustainability labels that effectively promote best practices are not prohibited. As such, sustainability labels based on best practices are stimulated, while those based on greenwashing sustainability labels are discouraged.

6.3 Limitations and Contributions

As it was found that predominantly private actors are involved in standard-setting, this analysis is part of an overarching question analysing the regulatory space of private actors to regulate sustainable development through decentred regulation. In practice, mainly EU competition law, EU consumer law, and EU fundamental rights law are inhibiting the potential of sustainability labels to promote sustainable development due to the noncritical application of normative assumptions derived from neo-classical law and economics. For that reason, these legal frameworks were examined. Theoretically, private actors operating in the EU may not only be subject to criteria underlying sustainability labels or the legislative provisions analysed. They may also be subject to many other VSS or legislative provisions such as EU company law, tax systems, product requirements for state aid or public procurement, manufacturing rules, and EU internal market legislation as stipulated by the fundamental freedoms Furthermore, international investment agreements (IIAs)[42] and international trade agreements (ITAs)[43] are also increasingly including sustainable development as an overarching objective. Further research could provide clarification as to how the objective of sustainable development could be aligned with the interpretation and application of these legislative provisions to private actors. Further

[42] See for the role of IIAs in fostering sustainable development, Alschner and Tuerk (2013).

[43] For example, the EU concluded an Economic Partnership Agreement (EPA) with the Cariforum group of States, which provides in Article 1 that its first objective is "to eradicate poverty and establish a trade partnership consistent with the objective of sustainable development". In Article 3, the Parties reaffirm 'that the objective of sustainable development is to be applied and integrated at every level of their economic partnership". See Economic Partnership Agreement between the Cariforum States and the European Community and Its Member States, signed on 15 October 2008, *Official Journal of the European Union* (30 October 2008), L 289/I/5, Article 1(a) and Article 3(1). See further on the objective of sustainable development in EPAs: Ruse-Khan (2010), p. 139.

research may be necessary to examine whether these legislative provisions can be interpreted and applied to private actors, such as private standard-setters and producers, to promote sustainable development[44]

Although further research may be required to get a broader understanding of how the objective of sustainable development could be aligned with the legal interpretation and application of legislative provisions to private actors, this book made important contributions to legal research and methodology, and legal and policy practice in the application and interpretation of EU law to sustainability labels. I will briefly discuss these contributions below.

6.3.1 Contributions to Legal Research and Methodology

The coexistence of EU law and sustainability labels results in interaction and competition for regulatory space.[45] Applying the analytical concept of "regulatory space" in the context of sustainable development is important as practice shows that the interpretation and application of EU law may inhibit the regulatory space of sustainability labels to promote sustainable development when legal reasoning is based on the noncritical application of normative assumptions derived from neoclassical theory. This is problematic as sustainable development is an overarching objective in EU law.[46]

To prevent EU law to inhibit the potential of sustainability labels to promote sustainable development due to the noncritical application of the normative assumptions associated with neoclassical theory, legal scholars need to get a firm grasp on how to apply relevant research from the social sciences to legal research and be aware of the assumptions built into the models of the social sciences used.[47] The case study analysis conducted within this book serves as a showcase as to how insights

[44] Such research has already partly been taken up by Ankersmit (2017), Chap. 5, n 36 and Partiti (2021).

[45] Ladeur (2011).

[46] See e.g. Preamble to Treaty on European Union (TEU); Article 3 TEU; Article 11 TFEU.

The preamble to the Treaty on European Union (TEU) states the determination "to promote economic and social progress for their peoples, taking into account the principle of sustainable development"; Article 3 TEU states that the Union's "aim is to promote (...) well-being of its peoples, which can arguably be equated with sustainable development or at least is an important component of it (see e.g. Stiglitz, Sen and Fitoussi, *Report by the Commission on the Measurement of Economic Performance and Social Progress*); the preamble to the Charter on Fundamental Rights (CFR) states that the EU "seeks to promote balanced and sustainable development; preamble to the 1994 Marrakesh Agreement Establishing the WTO recognising sustainable development as an overarching objective. The EU has mainstreamed sustainable development policies through the EU Sustainable Development Strategy, the Environmental Technologies Action Plan, and the Europe 2020 Strategy, for which sustainable consumption is a priority.

Article 2(2) SPS Agreement.

[47] Handler et al. (2005), p. 479, 489.

from the social sciences could be integrated into doctrinal analysis to align the objective of sustainable development with the interpretation and application of EU law to sustainability labels.

Integrating scientific insights from social sciences into legal analysis has the potential to improve legal interpretation and application by increasing the information base. In such a way legal decisions will be better informed, provided that the insights are used with the proper understanding of their strengths and weaknesses. A proper understanding and integration of different scientific insights from the social sciences are especially needed in a time where sustainable development has become an overarching aim in EU law,[48] while it is unclear how this aim can be operationalized.[49]

6.3.2 Contributions to Legal and Policy Practice

This book uses findings from social sciences to inform the doctrinal analysis in the interpretation and application of EU law to sustainability labels When properly applied, these findings are better equipped to inform the doctrinal analysis of misleading expression on food packaging and the anti-competitive effect of private sustainability initiatives then relying on a noncritical application of neoclassical law and economics.

More specifically, for the doctrinal analysis of misleading expressions on food packaging, this book shows how insights from cognitive science and behavioural economics could inform national authorities to improve sustainability labels to reduce consumer confusion. When national authorities only prohibit commercial information on food labels that gives a misleading impression to the *homo economicus*, companies may free-ride on sustainability labels that mislead the "real" consumer. This may have a negative effect on the potential of sustainability labels that stimulate best practices in sustainable production. Furthermore, when national authorities require mandatory disclaimers, while consumers are not able to adequately process such information, sustainable consumption may be deterred. National authorities are, therefore, advised to take insights from cognitive science and behavioural economics into account in the legal assessment of whether food

---

[48] Preamble to Treaty on European Union (TEU); Article 3 TEU; Article 11 TFEU.

See the preamble to the Treaty on European Union (TEU) stating the determination "to promote economic and social progress for their peoples, taking into account the principle of sustainable development"; Article 3 TEU states that the Union's "aim is to promote (...) well-being of its peoples, which can arguably be equated with sustainable development or at least is an important component of it (see e.g. Stiglitz, Sen and Fitoussi, *Report by the Commission on the Measurement of Economic Performance and Social Progress*); the preamble to the Charter on Fundamental Rights (CFR) states that the EU "seeks to promote balanced and sustainable development".

Article 2(2) SPS Agreement.

[49] Ruse-Khan (2010), n 43.

packaging gives a misleading impression to the "average consumer" as developed by the ECJ.

For assessing the anti-competitive effect of private sustainability initiatives, this book provides tools to economic regulators in the legal assessment of the proportionality of private sustainability agreements under EU competition law. Solely relying on doctrinal analysis or approaches derived from neoclassical law and economics may fail to accurately estimate the impact of such agreements on human well-being. This is problematic as human well-being is at the core of EU competition law. Additional insights from cognitive science, behavioural economics, as well as moral philosophy should be included in the legal assessment. In doing so, Member States are advised to clarify their stance on the normative assumptions and political questions underlying these social sciences approaches. Such a stance is necessary to use insights from social sciences as an addition to the information base in the legal assessment of the proportionality of private sustainability agreements.

The approaches proposed will enable public authorities to reduce the inhibitive effect of the application and interpretation of EU law to sustainability labels that emerged due to the noncritical application of neoclassical law and economics. When lawyers, judges, and economic regulators are willing to take the sustainable development objective of EU law seriously, they should integrate the relevant insights from social sciences into the interpretation and application of EU law when making normative assumptions.

References

Alexy R (2003) On balancing and subsumption. A structural comparison. Ratio Juris 16:433

Alschner W, Tuerk E (2013) The role of international investment agreements in fostering sustainable development. In: Baetens F (ed) Investment law within international law: integrationist perspectives. Cambridge University Press

Ankersmit LJ (2017) Green trade and fair trade in and with the EU. Cambridge University Press, Process-based measures within the EU legal order

Bernstein S, Hannah E (2008) Non-state global standard setting and the WTO: legitimacy and the need for regulatory space. J Int Econ Law 11:575

Botta M, Svetlicinii A (2013) The standard of Judicial review in EU competition law enforcement and its compatibility with the right to a fair trial under the EU charter of fundamental rights. In: Kerikmäe T (ed) Protecting human rights in the EU controversies and challenges of the charter of fundamental rights. Springer

Cafaggi F (2011) New foundations of transnational private regulation. J Law Soc 38:20

Cafaggi F, Iamiceli P (2013) Supply chains, contractual governance and certification regimes. Eur J Law Econ 37:131

Cashore B (2002) Legitimacy and the privatisation of governmental governance: how NSMD governance systems gain rule-making authority. Gov Int J Policy Adm Inst 15

Chernev A, Carpenter G (2001) The role of market efficiency intuitions in consumer choice: a case of compensatory inferences. J Mark Res 38:349

Cheyns E (2011) Multi-stakeholder initiatives for sustainable agriculture: limits of the 'inclusiveness' paradigm. In: Ponte S, Gibbon P, Vestergaard J (eds) Governing through standards: origins, drivers and limitations. Palgrave MacMillan

De Burcá G (1993) The Principles of Proportionality and its Application in EC law. Yearbook of European Law 13:105

De Vries S (2012) Consumer protection and the EU single market rules – the search for the 'paradigm consumer'. J Eur Consum Mark Law 4:228

Eden S et al (2008) Understanding and (dis)trusting food assurance schemes: consumer confidence and the 'knowledge fix'. J Rural Stud 24:1

Franck J-U, Purnhagen K (2014) Homo economicus, behavioural sciences, and economic regulation: on the concept of man in internal market regulation and its normative basis. In: Mathis K (ed) Law and economics in Europe: foundations and applications. Springer, Netherlands

Gerrits L, Edelenbos J (2004) Management of sediments through stakeholder involvement. J Soils Sediments 4:239

Golan E et al (2001) Economics of food labelling. J Consum Policy 24:117

Hancher L, Moran M (eds) (1989) Capitalism, culture and economic regulation. Clarendon Press, Oxford

Handler J et al (2005) A roundtable on new legal realism, microanalysis of institutions, and the new governance: exploring convergences and differences. Wis Law Rev pp 479:489

Harbo T (2010) The function of the proportionality principle in EU law. Eur Law J 16

Hatanaka M, Bain C, Busch L (2005) Third-party certification in the global agrifood system. Food Policy 30:354

Hoogland C, de Boer J, Boersema J (2007) Food and sustainability: do consumers recognize, understand and value on-package information on production standards? Appetite 49:47

Kahneman D (2003) A perspective on judgment and choice—mapping bounded rationality. Am Psychol 58:697

Ladeur K (2011) The state in international law. In: Joerges C, Falke J (eds) Karl Polanyi, globalisation and the potential of law in transnational markets. Hart, Oxford

Lange B (2003) Regulatory spaces and interactions: an introduction. Soc Legal Stud 12

Lipschutz R, Fogel C (2002) The emergence of private authority in global governance. In: Hall R, Biersteker T (eds) The emergence of private authority in global governance. Cambridge University Press

Marx A, Maertens M, Swinnen J (eds) (2012) Private standards and global governance: economics, legal and political perspectives. Edward Elgar

Mitchell R (2002) International environmental politics. In: Carlsnaes W, Risse T, Simmons B (eds) Handbook of international relations. Sage Publications

Padel S, Foster C (2005) Exploring the gap between attitudes and behavior: understanding why consumers buy or do not buy organic food. Br Food J 107:606

Partiti E (2021) Regulating transnational sustainability regimes under EU and WTO law. Cambridge University Press

Purnhagen K (2014) The Virtue of Cassis de Dijon 25 years later—it is not dead, it just smells funny. In: Purnhagen K, Rott P (eds) Varieties of European economic law and regulation. Springer

Purnhagen K, van Herpen E, van Kleef E (2016) The potential use of visual packaging elements as nudges. In: Mathis K, Tor A (eds) Nudging possibilities, limitations and applications in European law and economics. Springer

Ramus C, Montiel I (2004) When are corporate environmental policies a form of greenwashing? Bus Soc 44:377

Ruse-Khan GH (2010) A real partnership for development? Sustainable development as treaty objective in European economic partnership agreements and beyond. J Int Econ Law 13:139

Scott C (2001) Analysing regulatory space: fragmented resources and institutional design. Public Law 329

Smith P, Max-Neef M (2011) Economics unmasked: from power and greed to compassion and the common good. Green Books

Sperber D (1996) Explaining culture: a naturalistic approach. Wiley-Blackwell

Strack F, Deutsch R (2004) Reflective and impulsive determinants of social behavior. Pers Soc Psychol Rev 8:220
Stuyck J (2015) The Court of Justice and the unfair commercial practices directive. Common Mark Law Rev 52
Tridimas T (1999) Proportionality in community law: searching for the appropriate standard of scrutiny. In: Ellis E (ed) The principle of proportionality in the laws of Europe. Hart
Vandenbergh M (2007) The new wal-mart effect: the role of private contracting in global governance. UCLA Law Rev 54:913
Verbeke W (2005) Agriculture and the food industry in the information age. Eur Rev Agric Econ 32:347
Windholz E (2017) Governing through regulation public policy, regulation and the law. Routledge
Winn M, Angell L (2000) Towards a process model of corporate greening. Organ Stud 21:1119

Index

A
Amestoy, 92 F.3d 67 (2nd Cir. 1996), 13, 51, 70
Average consumer benchmark, 4, 12, 33, 34, 36–42, 49, 109

B
Behavioural economics, 5, 6, 11, 49, 102, 103, 108, 110, 112, 115, 116
Best practices, 6, 29, 30, 98, 101, 103, 104, 111–113, 115
Boggs, 622 F.3d 628 (6th Cir. 2010), 13, 51, 71

C
Capabilities approach, 83, 93–95
Case C-148/15 *Deutsche Parkinson Vereinigung* [2016] ECLI:EUC:2016: 776, 41, 87, 110, 112
Case C-195/14 *Teekanne* [2015] ECLI:EU: C:2015:361, 34, 36, 39, 40
Case C-309/99 *Wouters* [2002] ECLI:EU: C:2002:98, 85, 86
Case C-333/14 *Scotch Whiskey* [2015] ECLI: EU:C:2015:845, 41, 87, 110, 112
Case C-544/10 *Weintor* [2012] ECLI:EU: C:2012:526, 13, 51, 63, 64
Certifications, 12, 18–23, 25, 26, 28, 33, 104, 105
Charter of Fundamental Rights of the European Union (CFR), 12, 13, 51, 53–55, 57–60, 64, 65, 74, 107, 108, 111, 112, 115
Chicken of Tomorrow, 2, 3
Cognitive science, 5, 6, 11, 102, 103, 110, 112, 113, 115, 116
Competition laws, viii, 2, 3, 6, 11, 13, 14, 62, 81–98, 103, 106, 108, 109, 111–113, 116
Consumer confusion, 5, 68, 71, 72, 115
Consumer fair share, 3
Consumer law, 2, 6, 11, 12, 14, 33–49, 102, 103, 106, 108, 109, 112, 113
Consumer welfare, 11, 82, 87–89, 111
Contextual interpretation, 8–10
Corporate speech, 75
Cost-benefit analysis (CBA), 3, 83, 85, 87, 90–93, 95, 97
Credence attributes, 1
Critical legal analysis, 8, 101

D
Doctrinal analysis, 7, 11, 106, 112–113, 115, 116
Dutch Advertising Code Committee (ACC), 4
Dutch Competition Authority (ACM), 2–4

E
Effective interpretation, 10
European Commission, 13, 20, 58, 81, 112
European Convention on Human Rights (ECHR), 54–59, 61, 62, 65
Evolutionary interpretation, 10
Experience attributes, 1

© The Author(s), under exclusive license to Springer Nature Switzerland AG 2022
E. van der Zee, *Sustainability Labels in the Shadow of the Law*, Studies in European Economic Law and Regulation 22, https://doi.org/10.1007/978-3-030-95802-2

F
Fair trade, 1, 12, 25, 28, 29, 42–44, 46, 47
First-mover disadvantages, 4
Freedom of expression, 5, 12, 13, 51–61, 64, 65, 73–75, 110
Free flow of information, 76–77
Free riding, 4, 5, 33, 106, 115
Fundamental rights law, 2, 5, 6, 11, 12, 14, 51–79, 102, 103, 106, 108–110, 112, 113

G
Genetic modification, 5
Greenwashing, 1, 2, 6, 101–103, 105, 106, 112, 113

H
Homo economicus, 2–5, 112, 115
Human well-being (HWB), viii, 83, 87, 90–98, 101, 107, 108, 111, 113, 116

I
Information asymmetry, 1, 2

M
Mandatory disclaimers, 5, 115
Market failure, 2, 5, 21
Moral philosophy, 6, 13, 90, 103, 111, 113, 116
Multi-stakeholder regulatory instruments, 104, 105

N
Neoclassical law and economics, 6, 13, 29, 101, 103, 115, 116
Nike Inc. v Marc Kasky, 123 S Ct 2554 (2003), 67
Non-governmental organizations (NGOs), 21, 22, 24–26, 28, 29, 83, 104, 105
Nussbaum, M., 94, 95, 97

O
Objective well-being, 87, 89, 93–95
Organic, 12, 21, 22, 24, 25, 27–29, 33–35, 41–48, 102, 107, 109

P
Pearson I, 164 F.3d 650 (D.C. Cir. 1999), rehearing den., 172 F.3d 72 (D.C. Cir. 1999), 13, 51
Pictograms, vii, 12, 33–37, 40–49, 105, 106, 109
Political speech, 67, 74
Potemkin attributes, 1, 18
Principle of effectiveness, 10
Private sustainability initiatives, 2–5, 13, 98, 115, 116
Proportionality principle, 109

R
Rational actor model, 5, 6
Regulations, vii, viii, 5, 12, 13, 22, 25, 27, 33–35, 51–53, 57, 59, 60, 64, 65, 69, 72, 75, 88, 89, 96, 101, 102, 107–109, 113
Regulatory spaces, 101–103, 106, 113, 114
Resourcist approach, 93–94

S
Search attributes, 1
Search costs, 76–78
Sector-specific regulatory instruments, 105
Self-regulation, 4, 22, 104, 105
Sen, A., 90, 94, 97, 115
Social sciences, 11, 13, 30, 101, 114–116
Subjective well-being (SWB), 87, 90–94
Sustainable consumption, 2–7, 11–13, 17, 18, 26, 29, 30, 49, 78, 98, 101, 103, 106, 108, 114, 115
Sustainable development, 3, 11–13, 18, 29, 63, 82, 84, 86, 88, 101, 103–106, 113–116
Sustainable eel fund, 4

T
Teleological interpretation, 8
Textual interpretation, 8, 9
Tobacco Advertising Directive, 56
Treaty on European Union (TEU), 53, 55, 111, 114, 115
Treaty on the Functioning of the European Union (TFEU), viii, 2, 3, 13, 59, 63, 64, 81–88, 90, 95–98, 107, 111, 112, 114, 115

U
Utilitarianism, 90

V
Vienna Convention on the Law of Treaties (VCLT), 8

W
Whitaker v Thompson, 48 F. Supp. 2d 1 (D.D.C. 2002), 13
Willingness to pay (WTP), 3, 27, 29, 44, 46, 88, 91

The manufacturer's authorised representative in the EU is Springer Nature Customer Service Centre GmbH, Europaplatz 3, 69115 Heidelberg, Germany. If you have any concerns regarding our products, please contact ProductSafety@springernature.com

Printed and bound by CPI Group (UK) Ltd, Croydon, CR0 4YY
30/09/2026
02225513-0004